BETWEEN MIND AND NATURE

# BETWEEN MIND AND NATURE

## A History of Psychology

Roger Smith

REAKTION BOOKS

Published by Reaktion Books Ltd
33 Great Sutton Street
London EC1V 0DX, UK

www.reaktionbooks.co.uk

First published 2013



Printed and bound in Great Britain
by TJ International, Padstow, Cornwall

British Library Cataloguing in Publication Data
Smith, Roger, 1945–
Between mind and nature: a history of psychology.
1. Psychology — History.
I. Title
150.9-dc23

ISBN 978 1 78023 098 6

# Contents

# Preface

This is a new history of psychology, drawing on half a century of research, from the perspective of a historian of science. It is a critical history in the sense that it looks at psychology 'from the outside': it understands psychological beliefs and activity historically and does not take a psychological way of thought for granted. The book is for everyone interested in human nature and in the relations of the sciences and the humanities. I also hope students and psychologists of all kinds will find stimulus here, though it is not a textbook. Its coverage is unusually full. I write distinctively about the variety of psychological activity and the intellectual and social worlds of which it has been part. The history of psychology covers a field without clear boundaries, and I try to do justice to this. It is possible, though, to read chapters separately.

The book has an origin in an earlier and larger study, *The Fontana* (or *Norton*) *History of the Human Sciences*, published some fifteen years ago and out of print. This new book is different, with a sharper focus on psychology and much new material. In places, I have rewritten and brought up to date earlier material where that best suited my purpose; and I have also rewritten material taken from a version of the Fontana/Norton history translated and published in Russia.

I keep references to a minimum, listing sources for quotations at the end of the book. In a note on reading at the end of each chapter I suggest some material (mostly relatively recent), not including primary sources, which I find interesting, and where absolutely necessary, I indicate sources important for the discussion. As psychology has always been an international undertaking, I give titles of primary sources in the original language, along with a translation or the title of publication in English. As for the names of institutions or positions, I use capitals conservatively (names are often not in English or have changed over the years).

A book like this is really a collective work, albeit written in one voice. I build on the scholarship of many people over many years. It is my business to draw on a multiplicity of published work as well as on my own thinking, and it is simply not realistic to list sources comprehensively (it would be a book in itself). But I would here like gratefully to acknowledge my debt. I hope this book shapes collective scholarship into a larger picture. In England, Alan Collins, Jim Good, Graham Richards and Sonu Shamdasani have especially supported me in such a project, along with Ben Hayes at Reaktion Books, who launched it. I have written the book in Moscow, after taking early retirement from the history department at Lancaster University in the UK, and a little bit of the Russian influence, not least that of Irina Sirotkina, is evident, in addition to all the support which Lancaster gave me earlier for teaching and research.

CHAPTER ONE

# Early Strands of Mind

Because we are in the world, we are *condemned to meaning*, and we cannot do or say anything without its acquiring a name in history.
Maurice Merleau-Ponty

## UNFOLDING A STORY

How and why is psychology a taken-for-granted feature of contemporary life? Over the last 200 years or so in the Occident, the knowledge and desire to help and order people has taken psychological shape. Psychology is intrinsic to modern ways of life that value, seemingly 'naturally', the individual person, self-knowledge and self-control. Psychology is the vessel of very old longings for meaning and the vehicle for very modern self-fashioning.

As there is much reason to reflect, though, psychology has multiple faces. This book describes the multitudinous forms of psychology, as both scientific research and practical activity, in the context of social events and in the light of philosophical questions about the 'being' in being human. It covers the years from the French Revolution and the early industrial revolution to the present, the period of large-scale commercial and urban development. During this age of modernity, many areas of social life and masses of individual people began to look to psychology for orientation and practical help. Professional and ordinary people alike started, indeed, to become psychologists. After 1800, university disciplines came into existence in the natural sciences and humanities in something like their contemporary form, including, mainly in the twentieth century, the psychological and social sciences.

A rich intellectual life, it scarcely needs saying, in which beliefs about human nature and conduct were important, was present long before 1800. It may even appear obvious that some kind of belief about mind has been present in all cultures around the world at all times. But we should be cautious. Any way we characterize people, using terms like human nature, human being, man and woman, race and ethnicity, mind and body, is weighted with meaning with a long history. If you incline, for example, to say that everyone makes assumptions about human nature, you should bear in mind the existence of cultures which relate human and animal in entirely non-Western ways, indeed which think of animals as humans, not humans as animals. Some will think this curious but marginal, because Western science has simply got it right. If you are interested in how people live, or if you think that natural science is not the answer to every question, however, then the very terms in which different people think about the world has great significance. This book is about psychological terms, even about the possibility of there being specifically psychological ways of thought. In this sense, it is a 'critical' history.

Whether and how far we should describe earlier beliefs and other beliefs as psychology is a large question. There is, of course, a long historical background to the psychologies we have, just as there are ways of thinking from around the world which might well have something to contribute to Western psychology. Nevertheless, I shall stick here to what all would agree is psychology, which is modern and Western – and it is a quite large enough, and diverse enough, field.

History and ethnology alike are routes to understanding what is other, unfamiliar, and hence they are ways to create perspective: an ability to see ourselves as having a historical place. I think of history-writing as balancing two goals. Historians seek to be objective in attending to the record of the past, though they are aware anything said about the past is the achievement of their own thought in the present. This first goal involves not just looking at the archive but studying the context of thought and events, so that it becomes possible to see what happened as others understood it, not just as we understand it. The context is the local world – the world of institutions, people, cultural habits, language, intellectual assumptions, material resources, politics and so on – which makes it meaningful and important to think and do one thing rather than another. Contexts change with time and they vary with place. The second goal is to make sense: we want to know about what happened so that we exclaim, 'Ah! Now I understand!' History is one of the great arts of seeking to make the world intelligible and enjoyable.

Many books about the history of psychology like to begin with Aristotle, making it appear as if modern psychology is naturally heir to ancient wisdom. The rhetorical, indeed ideological, uses of this are obvious: psychology emerges out of the past as truth, not as one among other possible ways of knowing and living. The sense in which there may and may not be historical continuity with distant times is actually very difficult to unravel. I shall not attempt it here but instead discuss relatively recent times when there have been activities actually called 'psychology'. (As I will explain, the word was in use before the last 200 years, though not commonly in English.) I think there are large differences between attributing human actions to a soul, as Aristotle or Plato did (in different ways), and attributing them to a body, as neuroscientists now do, and I do not want to lump everything together as 'psychology'. In conventional accounts, a story about the rise of science links ancient and modern times: modern knowledge grows on the back of earlier beliefs and supplants them by confronting them with the facts of nature. The story is strikingly imperialist as it makes truth (or the best approximation to it) the possession of modern natural scientists.

I seek an alternative story. This first chapter explains the general thoughts which led to this, and it then sketches the historical background. Two chapters follow that describe the intellectual and social consolidation of scientific psychology in the nineteenth century. Chapter Two describes the major intellectual innovations which tied mind to nature: mental life to the brain and human nature to evolutionary history. The claim that human beings originate in nature, more than any other empirical claim, justified belief in the possibility of a science of human nature and hence a natural science of psychology. A number of people placed high hopes, just as they do now, on the turn to research about the physical world, especially the nervous system, in order to understand mind. But there were also other visions of the subject-matter of psychology. By 1900, as discussed in chapter Three, scientific psychology had a distinctive presence, most emphatically in Germany and the United States and less well delineated and on a smaller scale in France, Italy, Britain, Russia and other European countries.

The discussion of the twentieth century begins with a view of psychology going beyond academic disciplines and scientific research. Chapter Four traces the way people have come to think about managing individual and social problems with psychological knowledge and techniques. Much psychology is practical rather than theoretical or systematic, public rather than specialist. Chapter Five discusses leading ways in which scientists have claimed to make psychology into objective science. There have been a multitude of claims, sometimes varying country by country; it is necessary to make sense of why this happened. Chapter Six turns to the large role of ideas of the unconscious mind, in Freud's psychoanalysis but also in Jung's analytical psychology and in other practices. A century of irrational politics, it would seem, was also a century of attempts to come to terms with the sources of unreason in human nature. Chapter Seven describes social psychology, psychology focused on the relation between the individual, what the individual thinks, feels and does, and the social worlds of which each and every person is part. People, even solitary individuals, are social beings. A short discussion of Soviet psychology finds a place here because, in theory, the Soviet Union was a pioneering experiment in the truth of the social constitution of each person.

The final chapter describes psychology in the last 60 years or so and tries to link the past with the present. It touches on a huge field, by common consent not amenable to comprehensive overview. I consider a number of important debates, especially about how much human nature is due to biology and how much to culture and about whether mind is a function of brain (and psychology a branch of neuroscience). Everyone will be aware of the rise to prominence of the neurosciences, so I shall try to put this in perspective. Whether the neurosciences or my more sceptical views can create possibilities for a humane and intelligible psychology is an open question. I want readers to be able to see where their own experiences, views about psychology and beliefs about the identity of being human fit into history. This kind of history is like biography: it tells how and why we have become what we are. If we have lost the soul, what have we lost, and gained? If there still is soul, what is it?

If you do not like reflections on knowledge and want to go straight to the history, jump the next section.

## WHY HISTORY, WHICH HISTORY?

It appears self-evident: to understand a person, to understand human nature, you must learn some psychology. Nietzsche's declaration that 'psychology is now again the path to the fundamental problems' resonates (though there is a need to ask what he had in mind when he referred to psychology and why he thought it revived earlier insight). Bookshops and the media flaunt guides to the psychology of every human trait; students flock to psychology courses; worldwide, hundreds of thousands of people publicly claim the status of psychologist. It was not always so. Much of what we call psychology is very recent.

'Psychology' is a family name for a bewildering range of beliefs and occupations. There are people who test loss of function after injury and there are those who model face recognition; there are healers of the soul and there are public relations advisers to politicians. Many would think, however, that there is, or should be, a core of knowledge, the basis of a unified science. Indeed, used in the singular, the word 'psychology' flags this ideal. But what is the core subject-matter of psychology: mind, soul, behaviour, the brain, personality, discourse, mental structure or something else? The candidacy of evolutionary neuroscience, studying mental processes as adaptive functions of brains, currently has enthusiastic support as the unifying approach. By contrast, as the philosopher Gilbert Ryle observed, it may be that '"psychology" can quite conveniently be used to denote a partly fortuitous federation of inquiries and techniques' which 'neither has, nor needs, a logically trim statement of programme'.

I try to make some sense of variety through history. There is also an intrinsic fascination in understanding how people think about themselves, how they see their spiritual and material nature. Knowing how people in the past have thought, like knowing how people in other places think, also enlightens us. The historian Michel de Certeau commented, 'we travel abroad to discover in distant lands something whose presence at home has become unrecognizable.' A history, then, is a way of making sense by telling a story about why we live the way we do, why people shape the world psychologically.

It is common to explain the fact of much of psychology's history being recent in a simple way: psychology became a *science* only in the late nineteenth century. This makes psychology becoming a science the plot of the history of psychology. For much of the twentieth century, psychologists pictured this step as the adoption of the scientific method of objective observation in controlled experiments or testing. Recent decades have often seen the claim that psychology has become a science by becoming the biology of brain and behaviour. If we look more closely, however, we will find it difficult to maintain a historical story with one plot like this.

The history of psychology is inseparable from the argument about the way forward for psychology, the argument about what to expect from psychological knowledge and know-how. Indeed, I think history, if taken seriously and not just exploited for denigration or celebration, is one of the best resources we have for exploring not just the future of psychology but the future of human self-knowledge. History is for open minds. For closed minds, history is merely decorative or, more sinisterly, as with myths of nation formation, an ideological instrument to support one questionable view of the present.

There are a number of reasons why the history of psychology is a complex story and why psychology has many plots. When we study psychology we study ourselves – we are the psychological *subjects* we study as the *objects* of psychological research. How, then, is it possible for people to have objective knowledge of people? It would seem to be much easier to have objective knowledge about the stars or animals than about the subjective consciousness commonly thought intrinsic to a person and, naturally enough, references to subjectivity in science have negative connotations. In response, much of psychology has taken as its subject-matter not the subjective world of individual thoughts and feelings but something observable from the outside, like animals, or behaviour or brain; and psychologists have examined other people (like the mentally ill, or children or ethnic groups) but have been less at ease in addressing self, consciousness or subjectivity, let alone their own subjectivity. All the same, the large public for psychological ideas has obstinately looked to the field for answers to big questions. Is there an authentic self? Are my subjective feelings like those of other people? How does mind relate to brain? Is our life determined by a god, fate, genes, family conditions, moral personality? Can I exercise self-control? Why love? Why do people, individually and collectively, differ, and does it matter? What about the soul? Such questions bring an almost unmanageable complexity to psychology and its history. But they do make it significant.

Even if we declare the history of psychology to be about the field becoming a science, what kind of science is it? Science is a family of many different practices. In the English-speaking world, a science is (or is modelled on) a natural science; in continental Europe, a science (in German, *Wissenschaft*) is a body of knowledge grounded on rational principles and thought to be true – even theology may be a science. In English, 'science' earlier had this meaning too. In historical actuality, people have claimed to turn psychology into science in different ways. This is very interesting; but it makes it difficult to tell one story about the rise of scientific psychology. Of course, a contemporary psychologist who is quite sure psychology is, say, simply a branch of biology will not have this difficulty. But in fact, lots of people in the past and now think there is more in the picture than biology.

*There is no one discipline of psychology.* There is a huge variety of activities, some to do with scientific research but many others to do with practical life. Accurately speaking, there are psychologies in the plural and no such thing as psychology in the singular. There have been striking local and national variations.

Two large, linked questions immediately follow. One I have already asked: does psychology have ancient roots and should a history begin with Greek or even earlier wisdom? And, closely related, are there forms of psychology not based on Western science? In part, answers to these questions depend on definition. If by psychology one means any kind of belief (covering soul, spirit, mind, body, behaviour or whatever) about the nature of individual people, then, I suppose, we might think of psychology as universal. Even then, we must face the fact that different peoples do not so clearly differentiate either the individual or people and animals. Assumptions about what 'all people' feel or think abound in ordinary speech, but they may not be true. Less vague, and I think more correct, usage defines psychology as modern and Western. This usage does not devalue other ways of thought; rather, it allows them to differ – and indeed have value precisely because they are not psychological ways of thought. This is not an idle matter, since there is contemporary concern about the indigenization of psychology, the uptake by Asian, African, ex-Soviet and Latin American countries of the Western (in practice, overwhelmingly American) science and occupation of psychology. Should those who are active locally, those on 'the periphery', seek to acquire expertise from 'the centre' in the U.S.; or should they search in local culture for resources to obtain what they need to respond to local conditions? If they take the former course, their business is to catch up and contribute to a global project. If they take the latter course, will they still be contributing to psychology understood as a universal science of human nature?

There is yet another large reason for seeing complexity in the history of psychology: psychology is part of everyday social life. I want to explain this apparently banal statement – it is far from banal. The most obvious point is that the boundary between so-called scientific psychology and so-called popular psychology has always been very blurred. There is anxiety about the extent to which popular psychology (for example, in self-help guides) threatens to displace science in public esteem; but this is not new. Many of the roots of scientific psychology were in everyday life, growing out of knowledge of such things as temperament, the resemblance of children to parents and the expression of feelings. In the twentieth century it became a commonplace to assert, as an army chaplain, F. R. Barry, wrote in 1923, 'we are all psychologists today'. Psychological ways of thought have, of course, been unevenly distributed. But where we find them, we find what the historian Mathew Thomson has discussed in his book *Psychological Subjects*: people who think about identity and respond to life's challenges in psychological terms. A second point is that psychology has often advanced through application rather than through study for its own sake. Indeed, the pure-applied distinction is of little value in psychology, and it is more helpful, as Jeroen Jansz and Peter van Drunen have suggested, to refer to scientific and practical psychology, pointing to occupational rather than epistemological differences (differences in what people do, as opposed to differences in the grounds for the knowledge they have). The sociologist Nikolas Rose has well shown how much of modern psychology in Britain and the United States developed in practical settings – in schools, hospitals, prisons and in the institutions of social administration generally. In psychology it has not been the case that science advances and application follows; rather, the search to manage problems has generated science. A third point is that because people are both people who know and people who are the subject of that knowing, the development of knowledge, or new kinds of practice, changes people. There is a reflexive circle in psychology: new knowledge changes the subject-matter of the knowledge. The philosopher of science Ian Hacking described this as 'looping'; Michel Foucault noted that with human beings, the subject does not stand still: 'Whatever [thought] touches it immediately causes to move'. We become, it would seem, what we think we are.

This last point needs more discussion. A familiar model for what I am describing is psychotherapy, in this regard, comparable with moral and spiritual change. Through talk, discipline and ritual (and talk is also discipline and ritual), a person cultivates new thoughts and feelings and thereby tries to become, at least to a degree, a different person. The same kind of 'looping' is at work as belief and action intertwine in understanding phenomena, or difficulties, like depression, child abuse, autism and, earlier, multiple personality disorder. If it is polemical to say that particular ways of life, in some complex manner, create such conditions, there can be no doubt that belief that there are such conditions encourages the appearance of the conditions. These can be thought of as strong and weak versions of the reflexive claim. In either the strong or the weak version, the *history* of what has happened becomes part of what we need to know in order to understand why we have the psychological states we do. At the very least, we know that the expression of states like anger and shyness are historically and culturally variable. But we can go further than this and claim that practical life has generated psychological subjects – people with the kinds of psychological states we study as psychology.

There is a sense, I want to maintain, in which human beings, being human, have created their history. Reflecting in mind, for which language or a symbol system is a precondition, we have formed who we are. Telling the story of this self-creation will lead to a markedly different history from the story of psychology as the advance towards objective biological knowledge. If there is truth in this reflexive view, then knowledge of the history of how humanity has interpreted itself is knowledge of humanity.

The psychologist and historian of psychology Graham Richards suggested that it would help to think about these issues if we distinguish between Psychology (big P) and psychology (little p). The former is the science and the occupation: the psychologists, institutions, books and knowledge; the latter is the states, the processes, which the science and occupation studies and works on. There are Psychological studies of anger, while an angry person is in a psychological state. The strong reflexive claim is that as Psychology changes, psychology changes also; and vice versa. Thus, for example, I think it plausible to say that the spread of psychoanalysis caused people to develop identities, a self, built around their notion of unconscious forces. Reciprocally, modern ways of life fostered a turn to psychotherapy as a marketable approach to everyday problems. This, I want to be clear, is to state much more than the simple truism that the surrounding society continuously influences the development of the science (or sciences) of psychology.

It would be easy to be taken further afield with these arguments. I return, however, to my theme: the history of psychology is not and could not be a simple story. There was no one time when psychology began, no hero who started it all and no single line of development. There have been different claims about what a science of psychology is and there have been different psychologies. And psychological knowledge and practice have been factors in creating psychology's subject-matter. As a result, people will tell different stories depending on which kind of psychology they want their audience to know about. A contemporary history of psychology, authored by Wade Pickren and Alexandra Rutherford, gives considerably more space and attention to women and to African American psychologists than was the case earlier.

The existence of different histories, just like the existence of different psychologies, is not 'a problem' but the substance. The history of psychology is not about dull facts but a living debate over what sort of knowledge and practice we want.

## DESCARTES' WORLD

The Renaissance, beginning in Italy around 1350, and the early modern period, from about 1550 to 1700, saw huge changes. Paris, London and Naples began to have the character of modern cities; the printed book became common; kings and queens in France, Spain, Austria, England and Sweden, and later Russia, created centralized nation states; European voyages, especially to the Americas and to China, transformed European knowledge of people; the study of astronomy and motion, in the period between the major books of Copernicus (1543) and Newton (1687), replaced old views of the physical world with new science. There was excitement and amazement. When, for example, Europeans looked at indigenous American people, whom they called Indians, brought back for public exhibition, they saw people who, it appeared, had no clothes, no money, no property and no conception of the Christian God. Did this mean there was one human nature or many? What was the place of people without knowledge of Christ in God's world? Was it some people's natural place to be slaves?

Courts like those of Rudolph II in Prague and Elizabeth I in London, as well as wealthy merchants in towns such as Venice, Amsterdam and Augsburg, patronized an intense intellectual life and artistic culture. There was considerable interest in the practical arts of human affairs: in education, in using rhetoric and knowledge of human nature to command people, in the law and political thought, in medicine, and in new technology like printing, cartography, land surveying and financial accounting. The number of universities increased, especially in the lands of the Holy Roman Empire stretching across central Europe, and there was considerable interchange between them. There were also philosophies shaping novel ways of thinking about what makes a human being, and many later observers think this opened a road to scientific psychology.

No single figure has had more posthumous standing at the beginning of the new age than René Descartes (1596–1650). Even if scholarly work has shown that he had closer connections with late medieval philosophy than either he or his followers acknowledged, his work became, as it continues to be, a reference point for assessing how far new knowledge differs from old. Descartes himself was a characteristic intellectual of the first half of the seventeenth century, disturbed by what he thought a dangerous scepticism. The Protestant Reformation had begun a century earlier, and with it came profoundly divergent claims about the sources of authority and truth. The consequences may sound familiar: in a time of change, people fear uncertainties and search anxiously for absolute truth. It was Descartes' ambition to provide new and solid foundations for everything we can know. Very importantly, his model of certainty in reasoning came from mathematics. But whether certainty of the kind mathematical proof provides can be found in human affairs many people were to doubt, and, indeed, Descartes' own philosophy created new debate rather than a new synthesis of knowledge on being human.

Descartes lived as an independent scholar, much of the time in the Low Countries where there was greater freedom to publish than in his native France. Seeking the basis for certain knowledge, he concluded that we have knowledge of three kinds of being: of God, matter and mind. It illustrates his historically transitional position that he used cognates in

Latin and French for both 'soul' and 'mind': wanting to distance himself from Aristotelian notions of soul, he sometimes used the word 'mind', but he still used the word 'soul', as in his book on *Les Passions de l'âme* (1649; *The Passions of the Soul*), about self-control, to refer to what he thought a real entity. The natural world, including plants and animals, Descartes declared, consists of matter, extended substance in motion. It was an argument which made everything in nature the product of the size and motion of particles and hence subject to measurement and calculation. If nature can be quantified, it can be known precisely – just as Descartes' contemporary, Galileo Galilei, was proving in his new science of motion. In this way, Descartes provided the philosophical underpinning for modern physical science, the project to know the world in terms of causal sequences of moving matter (or, later, transformations of energy). Descartes and the new physical science removed what could not be measured, like soul, form and purpose, from the physical world; reference to such things had, according to the new way of thought, become unscientific.

Descartes framed all his thought in the light of certainty about the Christian God. He also self-consciously asserted his presence in the world, his own thinking, apart from matter. Like Montaigne in his essays, Descartes was fascinated by the authorial self. Knowledge, he presumed, must take into account the activity of knowing as well as what is known. Not able to doubt he was a thinking being, he claimed the world contains mind (or soul) as well as matter, and this mind, he held, is thinking substance, not extended in space, not in motion, not the sort of thing which is quantifiable. He therefore concluded that two kinds of entities exist: matter and mind, which have completely opposed natures, one characterized by mechanistic motion, the other by thought. This is his dualism – Cartesian dualism. In certain respects it was a clear way of thought. It laid out a programme for research in the physical sciences leading to the mathematical analysis of the physical world and to astonishing technology. But it was unbelievable.

Dualism was an unbelievable philosophy because it required belief in one place (at least) in the universe where matter and mind interact, even though they share nothing in common. This place is in each person. According to Descartes, human beings are unique because they are both body and mind. He thought that animals are machines, and he thought that much of human life, including, for example, passion, is caused by physical motions (in passion the blood is much agitated). Yet in humans, physical motions, such as those which take place in the optic nerves after light has fallen on the eyes, also interact with the thinking activity of the soul, and humans thereby have perception and imagination and reasoning. At once, as he perceived, this creates a conundrum: how do two completely different entities interact?

Over the long term, such thinking bequeathed to psychology a major problem. Descartes himself, we should note, never discussed anything called 'psychology', though he was interested in sensation, affection, memory and so on. When scientists later wanted to create psychological knowledge, they set about it with a mixture of knowledge of matter and knowledge of mind. Many problems of psychology and much of the interest in its history come from this. Wanting to know about themselves, people wanted to know about the place in the world where mind and body interact, precisely the point where Descartes, and much of science with him, was incoherent.

Dualistic thinking took many forms after Descartes. Naturally enough, there tended to be two opposed approaches: one, starting out from knowledge of matter and drawing mind into the material world; the other, starting out from mind, or soul, and making knowledge of matter secondary. As the range of positions was very great, though, it is good to be wary about talking as if there were simply two opposed institutionalized schools or traditions, empiricist and idealist. Further, insofar as there were empiricist and idealist tendencies, one side was *not* for science and against religion and the other side was *not* against science and for religion.

Historians of psychology, by and large, have had little to say about religion, mainly, one suspects, because of a tacit assumption that scientific psychology and religious belief are mutually exclusive, and the latter is therefore simply of no interest to a story about the rise of scientific psychology. For a number of reasons, this cannot be right. A priori, nothing sensible can be said about a conflict in general between 'science' and 'religion', since both words denote a huge cluster of activities rather than any one thing. The complex interplay of both empiricist and idealist tendencies in thought about science and religion illustrates this. Then there is the historical evidence for the large-scale involvement of religion in psychology – evidence for the general Judaeo-Christian background to thought on soul and mind and the improvement of 'man's estate', and evidence for specific influences, some of which I shall mention. When there were sharp differences, as of course there sometimes were, we can identify specific reasons.

In the eighteenth century, in contexts where knowledge, so to speak, ran in the direction from matter to mind, there were medical and experimental studies of the functions of the different parts of the body, including the brain and nervous system. By the end of the century, writers were linking reflex actions to the spinal cord and conscious sensation to the brain. The French physician, J. O. de La Mettrie (1709–1751), published a scandalous polemic, *L'Homme machine* (1747; *Man a Machine*), which denied the soul, and Denis Diderot (1713–1784), in witty and risqué essays, wrote about the play body and sex make with human life. The influence of theories of matter was also strong in British culture, and this had consequences for the shaping of nineteenth-century psychology. Locke and those who followed him, in order to understand the mind, supposed we should start with the sensations which originate in the physical world and which give us experience. This was to become the belief that the key to psychological knowledge is knowledge of sensation and subsequent learning, memory, behaviour and adaptation. We can follow the belief from the time of Kepler's work on the eye in the early seventeenth century, through nineteenth-century claims for the reflex basis of psychological events in the brain, to the U.S. behaviourist psychologists, to Pavlov and to modern cognitive scientists who construct computer models of mind. Psychology, from this viewpoint, studies the way a person responds to the world in which she or he is literally, physically, embedded. It is the point of view of the contemporary psychologist who thinks in biological, evolutionary terms.

Some histories of psychology have been written as if this were the only possible story. The reason is clear. The story vindicates belief that psychology is a rigorous science with the same kind of objective and systematic knowledge as the physical sciences. Given the success of the physical sciences in understanding and controlling nature, and the high status they have in universities and in society, this is extremely attractive to psychologists. It confirms the capacity of psychology, using scientific method, to make progress. Moreover, aligning psychology with the natural sciences holds out the prospect of the unification of knowledge, including knowledge of people, into one great scheme of knowledge of the natural world. This would seem to leave just the one 'hard problem' – the problem of consciousness – and the race is on to solve it following the way of thought which has proved so successful.

I shall, however, also attend to the contrasting approach, the approach to knowledge which, so to speak, has run in the direction from mind to matter. The symbolic birth of this tendency in argument, at least in modern times, is Descartes' famous claim that whatever he might doubt, he could not doubt that he had the capacity to doubt; in other words, he could not doubt that there is a mind thinking. Similar claims led numerous philosophers to start with the analysis of reason, rather than sensation, when they tried to say how it is possible to have knowledge. We can follow the arguments from the time of G. W. Leibniz (1646–1716), in the generation after Descartes, through Immanuel Kant (1724–1804), to Edmund Husserl (1859–1938) –

the founder of phenomenology – and his followers in the twentieth century. These arguments greatly influenced psychology, putting forward mental content, mental acts, language and the phenomenal world of consciousness (the world as it appears in subjective awareness) as the subject-matter of the field. The arguments are alive and well, if taking new forms, in psychologies which, in certain ways, have affinities with the humanities rather than the natural sciences. These psychologies do not generate causal knowledge of the kind found in the natural sciences, but their proponents would say that they nevertheless generate disciplined understanding, and hence, adopting continental European usage, also contribute to science.

The idealist tendency has often appealed to people who think that a human being is not a machine. Some writers, like Descartes himself, wanted to keep a place for a religious notion of the soul as a distinct entity, essential to being human. This remains the position of the Orthodox and Catholic faiths. Other writers, however, like the German philosopher Wilhelm Dilthey, were concerned with a science of psychology based on knowledge of what the mind, or in once common words, the human spirit, has achieved in language and culture; that is, they took the base of psychological knowledge to be the individual's historically specific activity. Arguing in this way did not necessarily presuppose the existence of the soul as a distinct entity but rather started out from the value-creating activity of humans in history. Yet others, following Husserl, like the Dutch psychologist F.J.J. Buytendijk, sought the basis of psychological knowledge in the immediate conscious world, in the feel and sense of significance in lived experience. Yet others turned to language (and this continues in the contemporary discursive psychology of Rom Harré), or to the archetypes of a collective unconscious (C. G. Jung), or to a constructive self (Carl Rogers) and so on. When modern neuroscientists, having identified the self with the brain, jump up and down and declare that they have disproved the soul, they skip over much of what matters in these questions. But this takes me to the conclusion, not the starting point, of the book.

## EIGHTEENTH-CENTURY PSYCHOLOGY

When Descartes wrote his philosophical books, the Latin word *psychologia* was already in use. This has directed historians, in recent studies, to a new understanding of the domains of knowledge relevant to psychology in the early modern period. 'Psychology', though, had no fixed referent in Descartes' age, and it certainly did not denote an institutionalized discipline. All the same, a number of teachers had begun to use the word to describe the study of the soul connected to the body, and in this we might see the beginnings of psychology understood as a natural science. I have already suggested, however, that we should not think of psychology as one thing with a specific origin.

The background is the late medieval and Renaissance study of Aristotle's text, *On the Soul* (*De anima*), and of commentaries on it, especially the theologically heretical commentary, from a Christian viewpoint, of Ibn Rushd (Averroes). This study spread as part of the preparation of students for higher levels of learning in law, medicine and especially theology. In the sixteenth century at the university in Marburg, influenced by the great Protestant reformer, the academic Philipp Melanchthon, and in Leiden, some teachers began to discuss the soul as part of the curriculum of *physica*, separating knowledge of the soul in relation to the body, which they discussed, from knowledge of the immortal soul, which they left to theologians. In this context there was reference to *psychologia*. To a degree, these teachers broke with the Aristotelian way of discussing the soul as the principle of life and, rather, adopting the model of Galenic medicine which related physiological functions to the body, treated mental activity as a function of the body. (Galen was the Roman physician of the first century in the Christian era whose writings formed the core of medicine for centuries.) In addition, the popular genre of texts on controlling the passions particularly emphasized the dependency of soul on body. Descartes took these trends further, separating rational mind, which he sometimes called 'mind' not 'soul', and the physiological activity of the soul tied to the body. During the seventeenth century, though *De anima* continued to be studied, many scholars' views became increasingly un-Aristotelian. There was a turn to empirical claims about human nature as part of a field called *anthropologia*. Sometimes a writer would refer to psychology as a branch of anthropology or even physiology; but there was also reference to *pneumatologia*, the study of spiritual entities (souls, angels), which could include psychology – here the name for the study of the rational soul (not connected to the body) and its immortality. Thus, if we wish to trace psychology in this period, it requires a broad view, a history of thought about human nature in general rather than an attempt to find something specifically like modern psychology.

Systematic reference to psychology as a domain of knowledge appears to have come after the writings of Christian Wolff (1679–1754), as least as far as the German-language world is concerned. Wolff, ranging across the whole of philosophical learning, laid out the relations of empirical and rational study of soul or mind as an academic field with a degree of independence from religious requirements. This further encouraged the study of anthropology, the study of Man as a natural being. (All eighteenth-century writers in English, indeed, virtually all writers till the 1970s, used 'Man' as the common generic term for human beings, and the capital letter was the norm before the last century.) Kant, for instance, lectured in anthropology for three decades at the university in Königsberg (now Kaliningrad, Russia), where he talked about all sorts of topics of interest to a general audience, like the nature of feeling for beauty and personal and national character, as well as cognition. The great *Encyclopédie*, edited by Diderot and d'Alembert (which began to appear in 1751), took up Wolff's treatment of the rational soul under the heading of 'psychology', and the heading thus entered French culture. Strikingly, the second, Swiss edition of the *Encyclopédie* (1770–76), treated psychology as a more substantial topic in its own right, and it cited the authority for this of Étienne Bonnot, abbé de Condillac (1714–1780) and the *genevois* Charles Bonnet (1720–1793). Bonnet wrote at length on *psychologie* and traced knowledge to sensory experience mediated by nerve fibres, supposing each fibre to have a predisposition to convey a particular sensation (as the strings of a musical instrument are tuned to different pitches). This might appear to have brought to a conclusion the potential of sixteenth-century thought to associate soul with brain not immortality, but in fact Bonnet was a Christian writer who, while relating what we can know empirically about the mind to nerve fibres, upheld belief in the immortal soul, even if we cannot have empirical knowledge of it. To understand Condillac, it is first necessary to turn to Locke.

John Locke (1632–1704) has a stellar position in histories of psychology, though he never used the word 'psychology', nor imagined he was contributing to a discipline separate from logic. His fame is that he wrote the canonical text displaying knowledge of mind beginning with sensory experience, and for those psychologists who identify themselves as natural scientists, *this* is the foundation of their field as objective research. Locke's own concern, however, was with how to distinguish secure knowledge from speculation, mere belief and enthusiasm, and his method was rational – analytic and attentive to language – rather than based on disciplined observation. Having, when young, experienced England torn by civil war, he ardently desired to know what distinguishes knowledge from enthusiasm, grounds for right action from grounds insecurely laid.

Descartes died in 1650, and in the following decades there was intense debate about his work. Isaac Newton published *Philosophiæ naturalis principia mathematica* (1687; *The Mathematical Principles of Natural Philosophy*), and, as people came to realize – few people could read the book itself – this demonstrated a new level of knowledge of the universe. Aristotelian thought about nature appeared completely inadequate in comparison. Locke, in *An Essay Concerning Human Understanding* (1690), which was begun many years earlier, undertook to explain how in theory it is possible to have the kind of knowledge which Newton had shown in practice we can have about nature. Locke concluded that there is only one route to reliable knowledge, through experience, just as Newton maintained that he had built on experience to reveal the laws underlying natural motions. Locke went on to describe in detail how sensation, and reflection on sensation, builds up complex mental life with thought and language. In addition, he promoted a hugely influential line of thinking, which treated the feelings as varieties of pleasure and pain, treated pleasure and pain as species of sensation and then explained our conduct or behaviour as a response to the pleasures and pains of sensory experience. He proposed the pleasure–pain principle: a principle which states that experience determines motives, and motives cause action, all in regular ways. The large consequence of this was the belief that experience determines character and conduct or, in modern terms, personality and behaviour.

Locke elaborated a theory of knowledge in which he used the metaphor of 'the blank slate', an image of the mind as empty before sensation inscribes something on it. He would have regretted using this image, if he could have foreseen the polemical use others were to make of it. Locke himself clearly attributed pre-sensory powers to the mind: the power to have sensations and the power to reflect, that is, the mind's capacity to judge about the ideas which experience provides; the mind also has the capacity of feeling. His image misled. Locke's theory of experience can look quite modern; but he did not question Christian faith, only distinguishing the kind of knowledge which faith provides from the kind of knowledge which experience provides – the type of argument which was to be a mainstay of modern attempts to assign religion and science to separate spheres.

The *Essay* was long and philosophical in manner. It nevertheless conveyed a striking and accessible picture of each person starting life as a young child with almost no mental life, and then, through experience and growing up, becoming an adult with complex thoughts and feelings. Locke tied what a person knows and does – who she or he is – to the experience which she has, the social world in which she grows up: a person's character is made not born. He thus assigned to knowledge of experience the most profound practical as well as philosophical importance. Over a century later, writers in English were to call this knowledge 'psychology'; for much of the eighteenth century, though, they usually wrote about knowledge of human nature. If we know human nature, they declared, we will learn how to educate people to build a better world. If we control experience, we will control action. This was 'Enlightenment', the word, as used by later historians, which gave a name to the age after Locke, the age of the *Encyclopédie* and the *Encyclopædia Britannica* (1768–71). Locke himself wrote a much read short guide to education, progressive for its time, in which he suggested ways to arouse a child's interest in experience and learning rather than crudely and cruelly beating knowledge into a child's memory. The new world, he and others anticipated, would be enlightened: once free of ignorance and religious prejudice, free from the vanity and greed of kings and tsars, people would use knowledge to control both physical nature and human nature, to reduce suffering and increase happiness. Abbé Gabriel Bonnot de Mably wrote: 'Let us study man as he is, in order to teach him to become what he should be.' It was the finest hope.

Locke used 'consciousness' in a way which made it central to English-language accounts of mind. It related to the Latin '*conscientia*' and the French Descartes used, '*conscience*', which is still current in French psychology; these words denoted conscientious thought, thought framed to conform to the moral value of truth. The Cambridge theologian Ralph Cudworth, however, not long before Locke published, had introduced 'consciousness' in order to characterize reflection on thinking as a *cognitive* ability rather than a *moral* act. 'Consciousness' and 'conscience', in English, came to parcel out cognitive and moral faculties; by contrast, the distinction between them did not come naturally in other languages. It may sound an esoteric story, but it paved the way for the kind of discussion which now takes place on the nature and origin of consciousness. For many modern scientists, consciousness is simply an empirical phenomenon, however puzzling, a fact, waiting to be explained. But the notion that consciousness is a fact is modern and questionable; it can be argued it is not a fact but a concept central to modern thought about humans as reflective agents. Moreover, the concept is linguistically variable. Thus consciousness is a modern concept, not a natural category, and there are other ways of discussing mind which do not pose consciousness as 'the hard problem' but an empirical problem, of scientific psychology in the manner in which modern neuroscientists understand it. Psychological puzzles come into existence in ways we can trace historically. If English language separates a normative (rule-directed) conscience from the phenomenal fact of consciousness, this is a historical and cultural achievement, part of a wider discourse separating normative, moral definitions of being human (to be human is to act according to certain standards) from factual claims about what is empirically the case in being human.

Going further into what he thought we can know, Locke discussed the self in a manner which shocked his contemporaries. Asking the question, 'What constitutes the We or I?', he answered that it is the continuity of consciousness over time. This was in marked contrast to the usual Christian attribution of the self to the unified soul. It is, Locke thought, the succession of our ideas and feelings, the continuity of consciousness, which gives people a self-identity. This suggested that if there is a break in continuity – even in sleep, as his critics immediately pointed out – the self is broken. Is a person then merely the sum of sensations and ideas which get replaced in new circumstances? Locke appears to anticipate the modern self, an aggregate of interior, changing particulars, a consciousness without conscience, open to the winds of change. In the eighteenth century, Laurence Sterne, in *Tristram Shandy* (1759–67), was to play gloriously with such thoughts. By then, it was indeed the novel which had become the vehicle for representations of human identity. The novel, as the medium par excellence for portraying the nature and causes of individual feeling and action, acquired a position of inestimable importance in forming a public attuned to psychological thought. Samuel Richardson's epistolary novel *Pamela; or, Virtue Rewarded* (1740–41), a sensation in educated Europe, was an extended drama of the subjective, and hence we might say psychological, struggles of the heroine to preserve her virginity, her self. The novel was *public*, integrating individual subjectivity in a collective psychological world. Locke, however, had already implied that identity might be unmade as well as made, vulnerable to circumstance as well as a creative force.

Along with the novel, letter writing, diaries and memoirs flourished, all contributing to forms of understanding in terms of individual, subjective and, as we would say, psychological identity. Other cultural movements also contributed, like the fashion for cutting silhouettes in black paper, making a profile of a person's appearance and thus outlining psychological character. Psychology thus originated in changes in ways of life and commonplace sensibility, not only in difficult texts.

Condillac in France and David Hartley (1705–1757) in England turned Locke's account of experience into detailed analyses of the way, as they thought, sensations combine to produce complex knowledge and conduct. (Language referred to 'conduct', a word with moral connotations, rather than to 'behaviour', a term characteristic of the modern attempt to separate fact from judgment, by observing only the outward form of the doing, about what a person does.) Condillac reused the model of the statue, which from a state of inert existence gradually acquires sensation through one sense after another. Taken up with one sensation, the statue exhibits attention; given a second sensation, the statue's awareness of the two gives memory of the first and comparison of sensations, the beginnings of thought; and so on. In spite of appearances, this picture of being human was not such a passive one, since Condillac imagined his statue in possession of desire or wants; it has an innate vitality to seek out sensation and to respond to it according to its pleasurable or painful qualities. As for human desires or wants, he linked their particular character to culture and circumstance. He formulated a whole, systematic course of instruction on his principles. Hartley, for his part, emphasized the association of ideas, the regular ways memory, thought and action, he claimed, depend on the juxtaposition of sensations in time and place and on their similarity and difference. Out of this the British utilitarian social philosophers, influenced by Jeremy Bentham (1748–1832), developed a political programme to create a world where the pleasures and pains of experience would lead people, naturally and inevitably, to act in ways which increase human happiness. James Mill (1773–1836), a follower of Bentham, in 1829 published his *Analysis of the Phenomena of the Human Mind*, in which he analysed – in an unflagging dry style – the different sensations, their combination, the accompanying feelings of pleasure and pain and the resulting conduct which, necessarily, pursues pleasure and avoids pain. The goal was a calculus of human conduct, the basis for a social and legal order which would make people seek the happiness of all. A contemporary young woman, Helen Bevington, with Bentham in mind, put her finger on the weakness of all this:

> They say he cherished men,
> Their happiness, and then
> Calmly assumed one could
> Devise cures for their good,
> Believing all men the same,
> And happiness their aim.
>
> He reckoned right and wrong
> By felicity – lifelong –
> And by such artless measure
> As the quantity of pleasure.
> For pain he had a plan,
> Absurd old gentleman.

We cannot, she supposed, measure happiness or virtue. Yet much modern policy, taking knowledge of individual human nature to be the basis for the good society, has these intellectual roots.

If we may question the empirical authority and practicality of utilitarian thought about the mind, we cannot doubt that it moved away from traditional Christian belief about the soul. Such thought did not cause the French Revolution, but the Enlightenment philosophers, the utilitarian social reformers and the actual revolutionaries all believed that it is possible for people to remake the human world. This depends, they understood, on human reason not, or at least not directly, on God. The constitution of the new republic of the United States of America found a voice for this belief that free men can take the future into their own hands. Yet even the United States excluded slaves and women from political society.

## THE SUBJECT-MATTER OF PSYCHOLOGY

Opposed to all this new thought were the conservative political powers of old Europe. Rulers from Madrid to St Petersburg maintained that social order depends on God-given hierarchy, absolute control from the top downwards and unconditional faith. Man is essentially fallen, conservative writers argued, and only Christ will redeem – not in this world but in the next. East of the Rhine, enlightened thought had little hold or took effect later. Medieval forms of social organization, in which the great mass of people remained illiterate, endured. In German, Austrian and Italian lands in the eighteenth century, and then in Russia in the nineteenth century, though, a stratum of people with education but without political representation emerged, and they longed for an enlightened future. In these conditions, psychological thought, focusing on subjective states, was an attractive means to assert individuality and freedom. Even in Britain and in the United States and France after their revolutions, where there was a turn towards economic and political emancipation, people countered social constraints by emphasizing psychological individuality. In German lands, an intense cultural life grew up around the many small courts, each with its associated university. The court of Saxe-Weimar-Eisenach was, in the late eighteenth century, home to J. W. von Goethe (1749–1832), who became a symbol of the ideals of German culture, and it directed the university of Jena, the town where, in the 1790s, the poet and playwright Friedrich Schiller, the philosopher J. G. Fichte, the Schlegel brothers and later the philosophers Schelling and Hegel, launched romanticism in literature and idealism in philosophy.

Whereas English, Scottish and French eighteenth-century thought, influenced by Locke, attended to the creation of mental life through sensation, German-language thought, influenced by Leibniz, stressed the activity of reason. By the second half of the century, there was a large literature on the different mental powers, which writers tended to group as faculties of reasoning, feeling and willing (the cognitive, affective and conative powers). This was the basis for two significant developments, and if one was abstract and at the margins of intelligibility, the other was concrete and at the heart of how ordinary, though modern, people pictured themselves.

Kant's critical philosophy examined what it is possible to know and the logical (not factual) conditions which make science possible. According to Kant, we necessarily reason about the world in a manner shaped by the logical structure of reason. Whatever the argument in philosophy, people found in Kant's writing support for something very much like a psychological argument: the mind works by actively creating frameworks, or structures, which shape experience, and certain structures may be innate. The general issues were exemplified in debate about vision: does the mind passively receive impressions from the eyes and in this way acquire knowledge about the spatially extended external world, or do innate mental structures shape what we see? (In addition, there was much interest in the role in vision of touch, including sensations of movement.) Many German-language writers concluded that psychological life is active, not a passive response to the world around: our world is indeed our world, since we have had a part in shaping what we know, feel and do. For many, it was a more realistic and attractive picture of the mind at work than the picture which they found in Locke and his followers and which derived everything from sensations – and treated feelings like sensations rather than transformations of self. Kant was himself much interested in such questions, though in his formal critiques he argued that a rigorous, quantifiable science of mind, comparable to mechanics, is not possible. He certainly wanted to advance psychological knowledge, which he discussed as a branch of anthropology, but he thought of such knowledge as closer to the practical knowledge of everyday life than to deductive science. The search for mental structures, which Kant did so much to foster, was to have a long history.

The second development was new enthusiasm for a culture of the self in psychological terms; for characterizing individual people in terms of subjective qualities and especially qualities 'of the heart', of feeling. The arts flourished as natural mediums for expressing this. The *Bildungsroman*, the novel of an individual character finding her or his place and purpose in the world, about the self acquiring shape and direction, came into its own.

Madame de Staël, in *Corinne; ou, l'Italie* (1807; *Corinne; or, Italy*), explored, with palpitating excitement, the extraordinary career and love of an ideal woman, both intellectually brilliant and sensitive to feeling and beauty, and linked this to ideals of nationhood. The expectation spread that there should be a science of the subjective self, a science of what really drives people from within. Romantic writers created an image of subjective identity founded on hidden and powerful feelings, and, at times, this pulled interest in the mind away from reason and away from a psychology modelled on natural science. Here are roots for much of the modern popular notion of psychology as the science of subjectivity. Here are roots of Freud's extraordinary influence as a psychologist in the twentieth century. In the early nineteenth century, however, it was more often literary writers, like Goethe or Stendhal, than people who called themselves psychologists, who introduced a large public to this way of thought. This was certainly the case, for example, in Russia, where the characters imagined by Pushkin, Lermontov and Gogol' provided lasting models of subjective life.

Brought up in conservative society and with religions suspicious of personal feelings, a romantic generation broke loose. It looked to the subjective self, to the psychological world of individual feelings, for the way to live truly, authentically and well. The artist, the person who finds in her or his own inner world the source of vivid expression for this reality, became a heroine or hero, a model for others about how best to live. As some historians have concluded, this, more than any development in science, may have drawn people away from traditional Christian faith, faith which turned to God rather than to the self for hope. Moreover, while an expression of emancipation for many individuals, romantic art was personally and politically ambivalent in its effects. If it focused on individual hope, it also emphasized individual despair. Romanticism expressed human problems as problems of individuals and their spirit, not problems of social structure and power, and, when it did inform social thought, it often fostered mythical beliefs about particular groups or nations, linking and dividing people by feeling and imagined ancestry, race and nationality.

The culture of the new psychology of the self was a public culture not a university culture. Some universities, indeed, like Oxford and Cambridge, were glorified men's clubs rather than hotbeds of learning, though in Scotland, by contrast, the universities were central to intellectual life. The universities changed in the nineteenth century and turned themselves into institutions committed to both teaching and research and established modern disciplines of scholarship in the natural sciences and what English writers call the 'humanities' – disciplines like philology, philosophy, history and the study of art and literature. Psychology, as also social science, had no position as a separate university discipline in the nineteenth century. Physiology, however, did establish itself as an experimental natural science, and in the first half of the nineteenth century there was considerable research on the nervous system, firmly consolidating knowledge that the brain is the organ of mind and reflex action the elementary unit of nervous function.

Writing about a subject which authors themselves called 'psychology' became more usual after about 1750, though this was common only later, not until the second quarter of the nineteenth century in the English-language world. The writing covered a very diverse field. There were studies of mental powers or faculties, analyses of sensation in general, experiments on colour vision, diaries about child development, discourses on individual moral character and accounts of feeling and the inner life. There was much awareness of people as social beings and of society as not just something which people happen to organize but fundamental to being human. For example, Adam Smith (1723–1790), then as now celebrated as the author of *An Inquiry into the Nature and Causes of the Wealth of Nations* (1776), professor of moral philosophy at the University of Glasgow, gave the central place in his account of human nature to sympathy, the individual capacity to feel what others feel. This, Smith thought, as well as being the cement of society, makes people intrinsically social beings. As Smith observed, other people are a kind of mirror to ourselves, and indeed he referred to the way we see and judge ourselves as involving our perception of an ideal 'impartial spectator'. The bridge between the so-called inner and outer worlds, many writers understood, is not unmediated sensation, and does not involve an isolated self seeing a physical world, but depends on experience mediated by *language* in a social world. We see and know by virtue of the linguistic culture in which we are a 'we'.

The study of language attracted a huge following. A discipline of philology, the background to modern linguistics, consolidated in the late eighteenth century, and there was much argument about what the history of language says about how people become the sort of people they are. It became common to believe that each person acquires a character and identity, just as each person acquires a language, as a result of living at a particular time and in a particular place. In the writings of J. G. Herder (1744–1803), Goethe's colleague in Weimar, this pointed towards social relations and cultural life, rather than individual minds, as the way to understand what people are and do. It encouraged research on the history of language, the rise and fall of civilizations and differences between groups of people

– groups increasingly linked during the course of the nineteenth century to political nations and to races.

In the eighteenth century, the fascination with human nature, whether focused on the association (a social metaphor) of ideas, sympathy or language, took this nature to be social. It is perhaps not too much to say that the roots of psychology lay in social psychology. Of course, 'social psychology', as a label for an area of psychological research, appeared only in the late nineteenth century. But it was studies of the situation of individuals in their social worlds, engaged in rational economic activity, exhibiting a moral sense, displaying national character or simply expressing sympathy in the family, which laid the foundations. Psychology was not always the study of individual capacities, conceived as prior to and independent of social life, which it was so often to become.

French writers, from early in the eighteenth century, divided the study of individual and collective actions, habits and customs, *le moral*, from the physical dimensions of human life, *la physique*. Examining *le moral*, Montesquieu wrote one of the most influential books of the century, *De l'esprit des lois* (1748; *The Spirit of the Laws*), in which he tried systematically to explain the different social, legal and political organization of different countries. Exemplary of the study of *la physique* was *L'Histoire naturelle de l'homme* (1749; *The Natural History of Man*) by George-Louis Leclerc, comte de Buffon, the introductory essay to many volumes of natural history which, among other things, compared different human types and the apes (the chimpanzee and orang-utan, since the gorilla was unknown in Europe until a century later). What we might think of as topics in psychology crossed the boundary between the moral and physical sides of human nature. When there was specific reference to *psychologie* (as in the *Encyclopédie* and Bonnet's writing), this focused on sensation. Then, in the 1790s, an intellectually influential movement known as *idéologie* attempted a unification of the moral and physical dimensions of human nature, showing how bodily life, sensory experience and social customs all interact to make people what they are. The term '*idéologie*' signalled a debt to Condillac's theory of knowledge or ideas. The aim of the *idéologues* was to provide a complete rational basis for social organization and, while the political realities of Napoleon's rule, followed by the royal Restoration in 1815, certainly excluded this, they had an impact, particularly through re-organizing medicine as a profession and lever of change in the social conditions of life. The physician P.-J.-G. Cabanis (1757–1808) drew the analysis of ideas into connection with knowledge of bodily processes, tying *le moral* to *la physique*, pointing towards an integrated medical and social science of the embodied person. Another physician influenced by *idéologie*, J.-M.-G. Itard (1774–1838), carried out one of the most famous of all experiments on human nature. The story is worth telling both for its own sake and because Itard tried to solve one of the main intellectual and practical problems subsequently to preoccupy psychologists and to shape their relationship with society: the problem of what a person owes to birth and what to upbringing.

During the cold winter of 1799–1800, local people saw a naked boy trying to steal food in a country area of southern France. Captured, it became clear that he had lived wild for some years, surviving by himself in nature. Here was a boy, the educated world thought, who was natural, without social training. He had no language, no cleanliness and liked to run on four legs. In Paris, Itard, who was involved in training deaf and dumb children, decided to demonstrate the effect of society on the development of the mind by showing the boy could learn to use language and behave like other people. Thus he began the experiment with the boy, who he called Victor, the wild boy of Aveyron. The boy proved a difficult pupil. After some years of intense teaching, Itard had to concede that, though Victor took baths and wore clothes, if reluctantly, he had not really mastered language. Itard's critics dismissed Victor as simply an idiot from birth. More thoughtfully, other observers wondered whether Victor had missed out on social contact at the point in a child's development when language learning would normally take place, and whether his incapacity showed that learning could not take place later. This experiment retains its intellectual interest and emotional impact – Victor was characterized in François Truffaut's beautiful film, *L'Enfant sauvage* (1970; *The Wild Child*) – and a number of stories about so-called wild children still circulate. What is it that makes a person a person and not an animal? What is natural if language and feelings require a certain kind of social development in order to exist? Much of later psychology set out to answer such questions; they proved far more complex than first thought.

At the same time as Itard was trying to educate Victor, Maine de Biran (1766–1824) began (he often did not complete) a series of essays which markedly deepened the philosophical and emotional sensitivity of *idéologie*. Revealing his links to romantic thought, his most extended and, in the long run, famous writing was his diary – a record of subjectivity. Maine de Biran re-examined sensation and found, underlying and preceding all experience, a feeling of active self, which he called *l'effort voulu* (willed effort). The unmediated, irreducible experience of the will of a distinct subjective self, he argued, must be the starting point for knowledge of mind. In the twentieth century, in the francophone world, this appeared to be the foundation stone for a conception of psychology based on the actual, existential subjective being of what we call a person: an intrinsically desiring and responsible being. It contributed a distinctive strand of argument sensitive to the will, to desire and to individual freedom in French philosophical psychology.

Maine de Biran influenced intellectuals in the nineteenth century but could hardly be said to have had a public impact. By contrast, Victor Cousin (1792–1867) and other teachers created an institutional identity, as the route to philosophy, for *psychologie* understood as the rational analysis of *la conscience*. This drew on Maine de Biran as well as a range of earlier writers, and though it hence earned the title of *éclectisme*, it established a notion of psychology as a specific field in France and taught this notion to generations of future teachers. Cousin initially occupied a somewhat liberal stance in French politics but, as conditions changed, his psychology looked more and more like a conservative Christian defence of the soul as the basis of all other thought. Opposition to it was to give rise to the notion of a 'new psychology' around 1880.

Though the opening decades of the nineteenth century were marked by idealism in philosophy in the German-speaking world, this was not incompatible with the empirical examination of consciousness – of what is in the mind – as a field with its own distinct questions and procedures. Kant's successor in Königsberg, J. F. Herbart (1776–1841), claimed to argue from the first, logical principles of philosophy to an account, confirmed by experience, of how the mind works. In *Psychologie als Wissenschaft* (1825; *Psychology as Science*), he described sensations and ideas as active forces competing with each other, the result of which constitutes conscious awareness under the shaping activity of the individual ego. Starting out from the logical principle of the unity of the soul, a principle he shared with the idealists, Herbart thought it possible to construct a quantified science of mental forces, a science analogous to mechanics. This science would, in its turn, he proposed, be the basis for *Staatswissenschaft*, the science of the state, and the basis for rational education. This made psychology a distinct field of knowledge and, like French *éclectisme*, made it a practical field central to the organization of education. A number of writers, like F. E. Beneke (1798–1854), followed Herbart, and though they are now little remembered, they did create an audience outside universities for a domain called psychology. Even earlier, F. A. Carus (1770–1807), professor of philosophy in Leipzig, had written the first history, *Geschichte der Psychologie* (1808; *History of Psychology*), in itself a significant influence in fostering the idea of a discipline of psychology.

All the while, first in Britain, then in France and Belgium, followed by the United States and Germany, the social world began the major transformation we know as the industrial revolution. This had its roots in population growth, the enlargement of commercial enterprise, the creation of new consumer markets and technological innovation. A science, political economy, with antecedents in government and military interest in the causes of a country's wealth and hence power, developed. The early theorists of political economy, including Smith, were preoccupied by the links between wealth creation, social organization and the character of people. Smith, who held the rather conservative view that natural sympathy is the cement of society as it has actually developed, along with the followers of Bentham, who wanted to reform conditions and hence reform people – a more radical politics – thought individual human nature the basis for thinking about social and political life. These writers did not refer to 'psychology', but they held beliefs about individual capacities, like sympathy and the motivation which results from pleasure and pain, which linked ideas, feeling and action. People in the educated classes, many of whom – like the Darwin and Wedgwood families in the English Midlands – acquired much wealth, took for granted the dependency of social progress on the cultivation of individual character. Thus both romantic sensibility and commercial society pointed to the importance of 'mental science' (a common nineteenth-century term), the science of the individual mind, for social progress. The coming of industrial modernity established individual mental character as both the drive and the moral purpose of progress, and romantic sensibility added moral and aesthetic value. This created a culture in which psychology flourished. The Scottish moralist Thomas Carlyle likened Bentham's and James Mill's theory of human nature to the machinery of the industrial revolution, calling the theory a mechanism to grind out human happiness. Influenced by German romanticism, he turned to genius as a model for the mind, thinking of genius as the capacity *to shape* the world in art or science or in politics or military conquest. Carlyle and the utilitarians alike, though, looked to the formation of individual character, to processes which became the subject-matter of psychology, as the basis for progress.

Whether under the label of psychology, anthropology, the science of man, mental science or moral philosophy, writers in the early nineteenth century studied mind and conduct in multiple ways. Many people, those with at least some education, sought new opportunities for themselves, their families and their communities, and they turned to literature and moral practices focused on learning, character, feeling and social relations.

Among intellectuals, some, like Maine de Biran and Cousin, advocated knowledge of the mind in order to understand subjectivity or matters of Christian faith. Others, like Bentham, demanded a science of human nature in order to control material and economic circumstances. There were real differences with social and political consequences, as John Stuart Mill (1806–1873), James Mill's son, discerned when he famously drew a contrast, in essays in 1838 and 1840, between the utilitarian reformer Bentham and the conservative poet Coleridge. Amidst all this argument, psychology, thus commonly named in English from about 1830, came to be a field for debating what it is to be human and for shaping the social world to make the truly human possible.

## READING

As many psychology students take a history of psychology course, there are numerous textbooks – which this book is not. Among these, T. H. Leahey, *A History of Psychology: Main Currents in Psychological Thought*, 6th edn (Upper Saddle River, NJ, 2003) covers the ground well in a traditional way, while J. Jansz and P. van Drunen, eds, *A Social History of Psychology* (Malden, MA, 2004) and W. E. Pickren and A. Rutherford, *A History of Modern Psychology in Context* (Hoboken, NJ, 2010) focus on the social relations of psychology, the latter with much institutional detail for the U.S. and a guide to reading. A different book for students, centred on key topics, critical, stimulating and historically informed, is G. Richards, *Putting Psychology in Its Place*, 3rd edn (London and New York, 2010). I include no illustrations; for this, see Jansz and van Drunen, and especially W. G. Bringmann, H. E. Lück, R. Miller and C. E. Early, eds, *A Pictorial History of Psychology* (Carol Stream, IL, 1997). I took a wide view, relating psychology to theories of human nature generally, bringing in more about the pre-1800 period and about the social sciences, in R. Smith, *The Fontana History of the Human Sciences* (London, 1997), also *The Norton History of the Human Sciences* (New York, 1997), which includes a large bibliography but of course no reference to more recent work. I put this history in the context of philosophical argument about understanding people in *Being Human: Historical Knowledge and the Creation of Human Nature* (Manchester and New York, 2007).

The new view of the early modern period is elaborated, with differences, in G. Hatfield, 'Remaking the Science of Mind: Psychology as Natural Science', in *Inventing Human Science: Eighteenth-Century Domains*, ed. C. Fox, R. Porter and R. Wokler (Berkeley, CA, 1995), pp. 184–231; P. Mengal, *La Naissance de la psychologie* (Paris, 2005); and F. Vidal, *The Sciences of the*

*Soul: The Early Modern Origins of Psychology*, trans. S. Brown (Chicago, 2011). For romantic sensibility and the self, C. Taylor, *Sources of the Self: The Making of the Modern Identity* (Cambridge, 1989), and J. Seigel, *The Idea of the Self: Thought and Experience in Western Europe since the Seventeenth Century* (Cambridge, 2005). I adapt my comments on 'consciousness' from papers by B. Hennig, 'Science, Conscience, Consciousness', and B. Carter, 'Ralph Cudworth and the Theological Origins of Consciousness', in special issue, 'History of the Science of Consciousness', *History of the Human Sciences*, XXIII/3 (2010), pp. 15–28 and 29–47. Also see J. P. Wright and P. Potter, eds, *Psyche and Soma: Physicians and Metaphysicians on the Mind–Body Problem from Antiquity to Enlightenment* (Oxford, 2000). C. F. Goodey, *A History of Intelligence and 'Intellectual Disability': The Shaping of Psychology in Early Modern Europe* (Farnham, Surrey, and Burlington, VT, 2011) is extremely challenging and its arguments yet to be assimilated. *Inventing Human Science* is a useful collection on the eighteenth century, while G. Richards, *Mental Machinery: The Origins and Consequences of Psychological Ideas, Part 1: 1600–1850* (London, 1992) contains much not easily found elsewhere. M. Billig, in *The Hidden Roots of Critical Psychology: Understanding the Impact of Locke, Shaftesbury and Reid* (London, 2008) shows in an original and accessible way that history can speak to the present.

Three English-language academic journals are especially relevant: *Journal of the History of the Behavioral Sciences* (founded 1965); *History of the Human Sciences* (founded 1988); *History of Psychology* (founded 1996).

CHAPTER TWO

# The Mind's Place in Nature

> I shall cut the frog open to see what goes on inside him, and then, since you and I are much the same except that we walk about on our hind legs, I shall know what's going on inside us too.
>
> Ivan Turgenev

## PHYSIOLOGY OF MIND

Origin myths create a sense of identity in scientific as in other communities. An oppressed nationality, a new occupation or a science without institutional standing all commonly claim a moment of birth and a founding father. This happened with psychology. The American psychologist Edwin G. Boring (1886–1968), defending psychology's status as an academic science, published an influential history, *A History of Experimental Psychology* (1929, revised 1950), with a father figure, Wilhelm Wundt, as its centrepiece. This German professor of philosophy, in 1879, equipped a small room in the university of Leipzig with instruments and created an experimental laboratory to train students for research in psychology. The symbolic act was using the experimental method in a special place made available for that purpose: this is what real scientists do. But the story perhaps symbolizes more than was intended, since much twentieth-century psychology became preoccupied with methods at the expense of knowledge. This myth portrayed psychology as the pursuit of science in universities. There was no hint that the social administration of modern life, new conceptions of people or the public business of self-help may, at the very least, have contributed.

If in 1800 there were people pursuing what they called psychology, they were few, and if they acted collectively, their stated purposes were many – anthropology, political economy, education, medicine and so on, as well as psychology itself. In 1900 by contrast, there were international congresses, laboratories, journals, taught courses in psychology and specialists claiming the name 'psychologist'. This and the following chapter follow this change.

In broad intellectual terms, the modern sciences of psychology depend on the argument that mind, or behaviour, cognition or whatever else is thought to be their subject-matter, is part of nature. This was not a new argument: the eighteenth-century search for the laws of human nature, whether in political thought or in medicine, tended in the same direction. It was, however, in the nineteenth century when two ways of thought, or principles, became established, which together continue to frame claims for psychology to be natural science: mind exists, always, as the accompaniment of nervous processes, and that mind, along with the body to which it belongs, is the outcome of evolution. Behind these principles was conviction in the uniformity of nature, in the operation of the same regularly acting causes across nature and the human sphere alike. This conviction does not exclude the possibility of humans having unique characteristics (such as language, religious beliefs), but it makes it an empirical, scientific matter to determine whether this is so, and if so, in what way. The conviction was not new, as some ancient belief about the soul had also drawn humans into continuity with nature, but the authority of the natural sciences for the conviction was.

This authority grew among public intellectuals and movements of thought, practical in intent, with considerable public participation. Universities and other elite learned institutions did play a role, but only at the end of the century was there an emphatic shift towards claims for specialist expertise on psychological topics. Even then, the public, 'amateur' interest did not just persist but expanded. The roots of psychology were in every walk of life, as we can see in the career of phrenology.

It took a long time for historians to take phrenology seriously, since it seems, to most people now, so obviously silly to believe in reading character from bumps on the head. Yet it was the founder of phrenology, F. J. Gall (1758–1828) who, perhaps more than anyone else, convinced people that the brain is the organ of mind; and it was public enthusiasm for phrenology which spread interest in the brain and belief that knowledge of the brain will make possible a great step forward in human self-help. Gall's initial question has intrigued people over and over again in an age of mass education: what causes one pupil to be cleverer than another or to have one ability rather than another? While still at school, Gall recalled, he had observed fellow pupils with protruding eyes who were especially gifted in the key ability needed by pupils, memory, and from this observation he developed his theory correlating strength of ability with the size of brain areas. As he also thought the enlargement of one part of the brain rather than another affects the shape of the head, it was possible 'to read' the bumps and reveal character. With this knowledge to hand, one could suggest career options or habits to cultivate or avoid so as to compensate for what nature had provided. He did not call his work 'psychology', but he started an influential movement to identify the internal nature of individual people, their brains and mental capacities, as the basis for external social relations.

Between 1810 and 1819, Gall, assisted by J. C. Spurzheim (1776–1832), published the four volumes and atlas of 100 engraved plates of the *Anatomie et physiologie du système nerveux* (*Anatomy and Physiology of the Nervous System*), a major contribution to the descriptive study of the brain. Gall collected skulls or casts – there are over 600 items kept at the Musée de l'Homme in Paris – as the empirical basis for the study of human character. He elaborated a language of physiognomy familiar to painters and sculptors and in everyday life (for example 'egghead' and the derogatory 'flat head'), and he interpreted busts of great historical figures. What was new, and a sign of the direction of science, was 'to found a doctrine on the functions of the brain' as a basis for 'a perfect knowledge of human nature'. Phrenology dignified gossip about individual character with objective status and made knowledge of the brain the guarantor of truth.

Gall's erstwhile collaborator, Spurzheim, travelled widely in Britain and the United States where, in the 1820s, he promoted phrenology as a guide to life and education. Knowledge of individual strengths and weaknesses, devotees argued, provides the information needed to mould character, including one's own. George Combe (1788–1858) carried the message to the lecture halls and lending libraries of provincial Britain. Prince Albert and Queen Victoria brought head-reading into the palace nursery. This spread a way of thought accepting that human nature is materially embodied in organs, the brain and skull, and is subject to research in natural science and to measurement. Phrenology, in its mild way, was a human technology, or, if you prefer, a practical psychology.

Gall believed in 27 basic faculties which determine innate capacities. Education, he supposed, can work with but barely modify an individual's given strengths and weaknesses. Spurzheim and other popularizers took a more optimistic and egalitarian view, and they advocated exercise as the way to strengthen or control a faculty. In their hands, phrenology became a science of human nature which makes it possible for everyone to help her or himself. This divergence of view among phrenologists was emblematic of a wider disagreement between those who emphasized innate endowment and those who emphasized education as the cause of differences of achievement. It was an argument that was to run and run.

Phrenologists described differences between individuals and between types of people (for example between men and women) on the basis of the claim that differences are ineradicably built into the physical fabric of people. This approach to the question of human difference, supporting much prejudiced thought about race and about men and women, was very characteristic of nineteenth-century empirical science, and often enough there was little to distinguish self-styled expert and public beliefs. It frequently implied determinism. Gall was certainly accused of this, though it is striking how popular phrenology, by contrast, propagated the values of individual learning and self-help. If there was a contradiction or at least ambivalence here, it was very widely shared. Throughout nineteenth-century thought, we find, as the German historian Leopold von Ranke observed, that 'freedom and necessity exist side by side.'

There is a curious twist to the history of phrenology in France. It was opposed from the beginning because of its apparent materialism – the Catholic physiologist Pierre Flourens (1794–1867) led the charge and reasserted the unity of the higher brain or cortex as the organ of the spiritual mind. He cited his experiments on removing parts of the brain in pigeons in support. By contrast, the would-be reformer of humanity under the banner of '*positivisme*', Auguste Comte (1798–1857), found in phrenology the positive science of facts about human character which he thought the basis for social policy. He specifically dismissed the possibility of psychology as a field (in his context, psychology meant the teaching of Cousin), holding it laughable, worthy of a comedy show, to force the mind to split into two so that one half could observe the other half. He called, rather, for objective scientists to study the brain, that is, be phrenologists, or study people in society, that is, be sociologists: they cannot study supposed subjective mental states in the middle.

Scientifically educated physiologists and doctors criticized phrenology from at least the 1830s, though it persisted as a commercially viable occupation into the early years of the twentieth century. There were other waves of enthusiasm, as for mesmerism, or animal magnetism, which swept through the public and encouraged belief in the physical embodiment of mental life. Following Anton Mesmer (1734–1815), mesmeric devotees used relations between a practitioner (usually a he), often enough a stage performer, and his subject (often a she) to show how people could do all sorts of things automatically, as it were, without conscious control. Whether attributed to a cosmic force or fluid (such as 'animal magnetism') or to suggestion, a psychological act (for which phenomenon the British surgeon James Braid coined the word 'hypnotism' in 1843), the enthusiasm brought mental life into close relation with the nerves. So did a large literature on the moral responsibility of individuals to help themselves to achieve health and moral character through knowledge of bodily physiology, and to become 'manly' or 'womanly'.

There are three large implications for the history of psychology in such activities: knowledge of the mind, psychology, was henceforth integrated with knowledge of the brain and nerves; individual and group differences were a major source of interest in psychology as domain of study; and psychology, from the start, was part of daily life as people sought to know themselves.

At this time, the formal study of mind in the English-speaking world was under the rubric of moral philosophy and, later, mental science. Thomas Brown (1778–1820), the holder of the moral philosophy chair at the University of Edinburgh, published his lectures to students on *The Philosophy of the Human Mind* (1820), which was the most used text in the field in the next decades. He systematically described the formation of the mind's content from sensations, though he declared that this is not a passive process but involves an active capacity to form the content by suggestion. Brown's or similar lectures in the liberal arts colleges in the U.S. provided several generations of young men with an education at once directed to their morals and to their self-understanding. College principals, like Thomas C. Upham (1789–1872) at Bowdoin in Maine, who published *The Elements of Mental Philosophy* (1831), spread a notion of psychology as moral philosophy. The situation was comparable with that in France, where Cousin led the teaching of *psychologie* as the discipline (in its double meaning as field of study and as training) in forming the minds and character of future teachers. Some of the followers of Herbart in German states, like Beneke, were also engaged in the same kind of moral-psychological occupation.

When proponents of the study of mind began to refer to mental science, they initially simply indicated that the field should be systematic, rational and well founded. Then, during the middle decades of the century, there was a spate of writings declaring that a true mental science requires recognition of the brain as the organ of mind and stating there was new knowledge to make this science possible. As an Edinburgh professor of medicine, Thomas Laycock (1812–1876), observed in 1860: 'All desires and motives are experienced in and act upon this important apparatus – and all are expressed by it; so that what the man is, in character and conduct, is the expression of the functions of this nervous system.' Physicians and physiologists across Europe expressed similar views. In Britain, two intellectual followers of J. S. Mill, Alexander Bain (1818–1903) and Herbert Spencer, also contributed large books, which to many later observers appeared clearly to delineate psychology as a field. Bain argued: 'the time has now come when many of the striking discoveries of Physiologists relative to the nervous system should find a recognised place in the Science of Mind.' Accordingly, he revised the earlier psychology of Hartley and of James Mill, which had analysed the content of mind in terms of associated ideas of sensory origin, both to relate mind to nervous physiology and to stress the place of activity, movement, in learning. George Henry Lewes (1817–1878), the novelist George Eliot's long-term partner and intellectual companion, wrote *The Physiology of Common Life* (1859–60), and something of the excitement about what he claimed for the embodied mind can be gauged from the long, drunken monologue at the beginning of Dostoevsky's *Prestuplenie i nakazanie* (1866; *Crime and Punishment*). Lewes's book, already in Russian translation, appeared to be the only education of a desperately poor family. For Dostoevsky, no doubt, this was a sign of the moral poverty to which abject material need led; for others, however, rendering the mind in material terms was a source of hope – it rendered human nature tractable.

Dostoevsky was very conscious of being in opposition. A large number of medical men were then promoting mental science, understood to be a science because it tied mind to body, as the foundation for self-knowledge, moral control and social administration. When the Victorian physicians in charge of the new asylums for the mentally ill organized themselves as a professional body and founded a journal, they called their publication *The Journal of Mental Science*. (It began life as *The Asylum Journal* in 1853, acquired the new title in 1858 and, in 1963, became *The British Journal of Psychiatry*.) In 1869, the parent body became the Medico-Psychological Association, a sign of the close and historically influential relations between psychology and the medical speciality concerned with mental illness. Only in the twentieth century was 'psychiatry' to become a common term and psychiatry and psychology to become occupations distinct because of training and methods. Even later, the division of labour and respective standing of psychiatry and medical, or clinical, psychology remained at issue. Many doctors persisted for a long time in believing in their own special psychological expertise; the neurologist Henry Head, who worked with soldiers injured in the First World War (and featured in Pat Barker's novels), certainly thought psychology part of his clinical and therapeutic field.

A number of Victorian physicians even took up the term 'mental physiology' to signal the direction which they thought psychology should take. The most forceful writer in this vein was Henry Maudsley (1835–1918), who made his career as a consultant about madness to the very rich: he delivered absolute discretion. He became publicly known for his mordant pessimism about the human failure he saw around him. However much such physicians emphasized the dependence of mind on brain and, increasingly, illness on heredity, they usually denied that they were materialist – an accusation which their critics tended to bandy about. Certainly, they intended to enhance not detract from the power of each person's will and responsibility. They were one and all zealous moralists, telling people what they should not do (not to drink, not to masturbate, not to waste women's precious and limited energy on learning) on the basis of what they claimed to know about physiological psychology. By the late Victorian period, the very triteness of the messages was a mark of the extent to which a psychology bound up with the life of the body had become taken for granted.

To understand this further requires a little more attention to what had actually been achieved in physiology. Experimental physiology was one of the natural science disciplines to become specialist enterprises in the German universities in the decades of the 1830s and 1840s. Then and later in the nineteenth century, physiology seized the imagination of young people, fed up with old prejudices, as the route by which knowledge would finally become objective and true to real human nature. By contrast, clerics and conservatives, some of whom were academics in the disciplines of history, philology and philosophy, looked on physiology's claims to elevate culture with suspicion; sometimes they simply attacked physiology as outright materialism. In German lands after the failed revolutions of 1848, or in Russia in the 1860s, with hopes, soon dashed, for a new political start, the arguments had a public face which challenged the very essence of being human.

Physiology, under the leadership in the 1830s of Johannes Müller (1801–1858) in Berlin, promoted the expert experimental study of humans as material systems. It was principally but not uniquely a German science, since a career structure for specialists developed in Germany before anywhere else. In Paris, François Magendie (1783–1855) began to carve out a distinct place for physiology in the medical curriculum in the 1830s. He was followed in the mid-century by Claude Bernard (1813–1878), who did brilliant studies of blood sugar regulation and of nervous control and argued forcibly for physiology as the foundation of scientific medicine. In Britain and the United States, as in France, medicine offered an early home to physiology, and hence to the study of the embodied mind, but physiological research on a large scale did not really get under way in these countries until the 1870s.

Physicians as a group were markedly sympathetic to material explanations of human nature, and public opinion supported academic medical activity even when there was some fear of its materialist implications. The great innovations in urban public health in the second half of the century consolidated links between medicine and government and encouraged the public to accept natural science ways of approaching human affairs. Many physicians believed in scientific medicine as the true science of man, and when the conviction grew that the disciplines of physiology, anatomy, histology (the science of tissues), physiological chemistry and – at the end of the century – bacteriology and biochemistry gave medicine a scientific base, a materialist science of man looked assured.

Magendie's experimental studies, in the early 1820s, on the separation of incoming sensory nerves and out-going motor nerves to and from the spinal cord, confirmed the structure of the nervous system as the organ linking sensation and movement. Then, in the 1830s, argument between the English physician Marshall Hall (1790–1857) and the German physiologist Müller resulted in the description of the reflex as the elementary unit of nervous function. This enabled scientists to describe an organism's achievement of purposive movement, for example a frog scratching itself, as the outcome of material causes, the structural relations of the nerves. It provided a model for explaining a mental attribute, being purposive, in physical terms. The concept of the reflex informed a research programme to determine the material correlates of mental processes, thus – this was the hope – recreating vague psychological descriptions as precise physiological questions. The research involved detailed experimental analysis of the conditions affecting the action of reflexes, often with the poor frog as the experimental subject. Scientists reported, for example, on reflexes after they had cut the spinal cord at different levels or altered the intensity of response with drugs (such as curare). In addition to this kind of work, which required training and created specialists, both physiologists and more general writers used reflex action theory to understand aspects of human life which appeared to take place automatically. Authors used the idea of reflex action to explain acquired habits and skills, sleepwalking, hypnotic performances, spiritualist and table-turning experiences, aspects of madness and much else besides. They created a public familiar with explanations of human life in terms of the underlying causal physiology of the body. Physiological psychology replaced moral philosophy as a route to knowledge of mind.

The decade of the 1840s was a watershed in the history of physiology as it was then that a group of brilliant students under Müller committed themselves to explaining life exclusively in physico-chemical terms. Emil Du Bois-Reymond, Hermann Helmholtz (1821–1894), Carl Ludwig and Ernst Brücke all went on to hold chairs at major universities and to shape the discipline of physiology over the next half-century. At the same time, the principle of the conservation of energy (the first law of thermodynamics, to which Helmholtz made a decisive contribution) gave great authority to the belief that only causal processes of a physical character could exist in nature. There could be no mental act 'from outside' adding to the energies of nature, no miraculous intervention of soul. For the many writers of popular science, this vindicated the claim for physiology to be the only route to a science of psychology. It was more and more difficult for those who thought otherwise to challenge the authority of natural science.

Physiology attracted a public audience because it engaged with what many people who were opposed to the old European political order considered the real material conditions of life. As in the eighteenth century, radicals, like the philosophers Ludwig Feuerbach and Karl Marx, found in the realities of the physical body and of physical labour the facts with which to attack idealist, Christian, conservative beliefs. Hopes for political change came to a head in 1848, with disastrous results for the radicals. One of the physiologists who contributed to radical thought with popular studies of physiology was Jakob Moleschott (1822–1893), a Dutchman, who taught at the university of Heidelberg. He memorably linked the subservient position of Europe's peoples (he was here writing about the Irish) to their diet: 'Sluggish potato blood, is it supposed to impart the power for labor to the muscles, and the enlivening verve of hope to the brain? Poor Ireland, whose poverty breeds poverty . . . You cannot win! For your diet awakens powerless despair, not enthusiasm, and only enthusiasm is able to blow over the giant [England] through whose veins courses the energy of rich blood.' When conservative regimes, trying to ensure 1848 would not happen again, repressed liberal as well as revolutionary demands, disillusioned people turned with renewed energy to natural science as the authority which, in the long term, would subvert the alliance of political reaction and religion. Moleschott's exclamation, 'Life is an exchange of matter', produced a conservative outburst that led to his resignation. He eventually took a post in Zurich.

Conservative power was an oppressive reality in tsarist Russia. But 1855 saw a thaw; it was a turning point for the development of the natural sciences as the basis for knowledge of people. The death of Nicholas I and fear of backwardness, symbolized by defeat in the Crimean War, ended an attempt to impose total control over higher education and restrict belief about human nature to the faith of the Russian Orthodox Church. Under Alexander II, even after hope faded for political liberalization, students travelled to Western Europe, mostly to German-language universities, for an education. Many idealistically believed that knowledge of the natural and the historical sciences would somehow itself precipitate reform. The word 'intelligentsia' came into use to describe this class in Russia: a class of unempowered but educated people who believed in objective rational knowledge, science, as the basis for civilized life. A total lack of representative government turned this class either into dissidents, sometimes into nihilists, or into scientists and doctors who focused their political activity on their profession and the modernization of the social administration of the state. Under the tsars, as later under the Soviet system, intellectual commitment to science, experienced as a source of authority which transcends political power, acquired intense personal significance as the means to sustain integrity under repressive conditions.

The journalist N. G. Chernyshevsky (1828–1889) wrote a long article, 'Antropologicheskii printsip v filosofii' (1860; 'The Anthropological Principle in Philosophy'), in which he described the objective, material conditions of being human. His formulation, though vague, was full of meaning to his readers as it diametrically opposed the conservative view which grounded human nature on a spiritual and immortal essence. Chernyshevsky, in this like Comte and Feuerbach, put forward humanity as the purpose of existence, and in a didactic novel, *Chto delat'?* (1863; *What Is to Be Done?*), written – however extraordinary it may seem – from a cell within the Peter and Paul fortress in St Petersburg, he gave his beliefs fictional expression. He chose two medical students among his protagonists, and he described them studying physiology and recognizing that sentimental love is a material necessity requiring honest openness to the forces of life. He portrayed them forging a new type of person, 'the new man'. He contrasted true human needs with the false needs produced by tsarist society, and he portrayed science as knowledge of true needs. The novel inspired a generation, and not least the young Lenin, with a vision of a science of man as a means of liberation. With conservative fears on the increase, however, in 1864 the government exiled Chernyshevsky to Siberia.

Among the Russian students to travel to Berlin, Leipzig, Heidelberg, Vienna, Zurich and Paris in the late 1850s, was Ivan Mikhailovich Sechenov (1829–1905). He returned to St Petersburg in 1860 and taught experimental physiology and contributed to the establishment of scientific medicine. He carried out experiments, very much on the German pattern, on the frog's reflexes, and he contributed to an international debate about inhibition in the regulation of movements. But he also had a larger ambition: to make this physiological work the basis for scientific psychology, an approach to the human mind which would be objective, as it would base itself on the material conditions of mind, that is, on the brain. He therefore wrote provocative essays for a general audience as well as scientific papers. The most famous of these essays, 'Refleksiy golovnogo mozga' ('Reflexes of the Brain', which appeared in its first version in 1863 but was not translated into French and English until 1884 and 1935 respectively), sought physiological analogues for psychological processes. His greatest challenge to orthodox (and Orthodox) belief was to develop the physiological concept of inhibition as an analogue for the will. Sechenov claimed that his theory did not detract from individual responsibility, which is what his critics, and the censors, feared; quite the opposite – only scientific knowledge, he thought, will make moral progress possible. Ivan Turgenev immortalized the essence of the debate – though Sechenov was not the source – in his novel, *Otsy i deti* (1862; *Fathers and Children*). The central character of the story, Bazarov, was a medical student and a radical materialist, a 'nihilist' in Turgenev's sense of the word, as he accepted nothing which could not be proved by material facts. He had a clear thesis about psychology, which gives this chapter its epigram. In Turgenev's story, however, Bazarov did not know what was hidden in himself: love, self-sacrifice and despair.

Soviet physiologists and historians claimed that Sechenov succeeded and founded a science of objective psychology. Sechenov himself, however, probably recognized just how large the gap was between knowledge of the brain and the experiential richness of mental life. Certainly, neither his nor similar projects to found a physiological science of mind translated from speculation into physiological detail in the nineteenth century. This was the general situation in mental science. Commenting on the hope that psychology would advance with knowledge of the brain as the organ of mind, Bain, as late as 1893, concluded: 'Introspection is still our main resort – the alpha and the omega of psychological inquiry: it is alone supreme, everything else subsidiary.' Many psychologists at the end of the century agreed, as we shall see.

The experimental physiology of the nervous system nevertheless became a major area of research, hand in hand with clinical studies of brain damage and disease, which formed the medical speciality known in the twentieth century as neurology. The development of antiseptic operative techniques and anaesthesia made possible new experimental studies of the brain. Among the most tangible results was the announcement by the German researcher E. Hitzig (1838–1907, working with G. T. Fritsch) in 1870, followed by the English physiologist David Ferrier (1843–1928) in 1873, that electrical stimulation of the cortex of the higher brain demonstrates the localization of function. It appeared possible to map sites in the brain where different functions, mental processes, like the control of speech, go on. Building on earlier work, a number of French doctors in the 1860s claimed, on the basis of studying the symptoms of patients and then looking in their brains after death, to locate speech centres. To superficial observation, it appeared as if Gall's theory of the localization of mental activity was right all along; but in fact Gall had localized mental faculties, while the new theories localized sensory-motor functions.

Best remembered now is the spectacular case of Phineas Gage. In 1848, during construction of a railway in Vermont, an explosive blast projected a tamping iron right through his cheek, under one eye and out through the top of his head; yet he lived. Enthusiasts for localization, then and now, tell the story of how he exhibited symptoms demonstrating loss of function in the damaged region of the fore-brain. But others now claim, partly on the basis of a photograph and of reconstructing how he must have looked, that he suffered psychological damage to his whole identity; certainly, for a while he exhibited himself publicly as a freak. Did he or did he not exhibit localized loss of function? Here we have dramatized a large argument: what is the balance of evidence between claims based on experiments with brains, as if we were dealing with a machine, and claims based on our everyday knowledge of the social nature of psychological performance after trauma?

Ferrier did his early work in a small room in the West Riding of Yorkshire Lunatic Asylum in Wakefield, which suggests something about the conditions for physiological research in Britain. He brought his studies together in a book on *The Functions of the Brain* (1876), where he filled a substantial chapter with comments on the mental function of the cerebrum or higher brain. His hopes for what physiology could do for mental science, or psychology, were encapsulated in his statement: 'thought . . . is in great measure carried on by internal speech.' (Sechenov had a similar notion.) This made the significant claim that scientists will be able to understand reason, traditionally held to be central to man's spiritual essence, as activity in the nervous pathways of sub-vocal speech. Speech, in this view, is an integration of sensory-motor acts which culminate in the co-ordination of muscles, rather than the exercise of social-linguistic rules by people. It appeared to enable scientists to study rational thought experimentally. But the dream of what science could achieve was one thing and its realization another. At the end of the century there was something of a backlash against 'brain mythology', and psychology developed other routes to become a science. Even within studies of brain, there was to be a long-running debate between those whom the French historian of science, Georges Canguilhem, called *les localisateurs* and *les totalisateurs*: those who began with division and those who began with wholeness.

## THE EVOLUTIONARY NARRATIVE

Public, medical and academic interest in the nervous system thus began to flourish in the nineteenth century, drawing mind inexorably into connection with the material conditions of life. This disturbed conservatives in politics and those who thought human purposes rest on transcendent qualities of faith, reason or the human spirit. As psychology developed into a distinguishable field of study and interest, it still retained its place as a contribution to 'ultimate' questions. Cheap comment, then or now, about science versus religion did little justice to the richness and variety of what was argued, though it was clear to everyone that decision about the way forward for a science of psychology could not be separated, in the last analysis, from philosophical questions, perhaps religious in content, about what the subject-matter of the field actually should be. Many psychologists have thought that empirical experience simply gives us, presents to us, the field's subject-matter. This, however, is unsustainable: each and every experience is 'an experience' in the light of pre-existing concepts, a classification of what exists and a language. Each and every experience has a history.

During the nineteenth century, it was evolutionary thought which focused public discussion of these issues. Along with the physiological approach to psychology, evolutionary ideas embedded study of mind in study of nature. Writing after the bicentenary of the year in which Charles Darwin (1809–1882) was born and the 150-year anniversary of *On the Origin of Species by Means of Natural Selection; or, the Preservation of Favoured Races in the Struggle for Life* (1859), there is a danger of overkill in talking about evolution. The evolutionary image of 'man's place in nature' (in T. H. Huxley's phrase) did, however, seize the public imagination to a special degree, and reference to Darwin in debate on science and religion has remained, and seems set to remain, the preferred trope. Further, no one will question the fundamental importance of evolutionary thought for psychology. I would, in summary, list the following: establishing authority for a science of psychology with the evidence that humans are part of nature, evidence so good that the onus of proof shifted to those who would deny it; pointing to function, and later behaviour, not mental content, as psychology's special subject-matter; emphatically reinforcing an interest in individual and group differences as the stuff on which the mechanism of natural selection operates; enhancing belief in the 'brute' origins of human nature, identifying the irrational and instinctive in motives; and consolidating the classification of human types and human history from primitive to civilized. This section will put some historical flesh on these bare bones. Beyond this, I inevitably have in mind the modern status of evolutionary theory, to which I return briefly in conclusion, as the foundation for a fully naturalistic approach to philosophical questions.

The story of how Darwin, having returned from his voyage around the world on HMS *Beagle* and familiar with the political economy of Thomas Malthus, arrived, in the late 1830s, at a theory of the origin of species by descent from previously existing species through a process of natural selection does not need to be retold. It is relevant that he delayed going public and that while he delayed there were a number of intellectual shifts, including the elaboration of physiological approaches in psychology. When Darwin did reach out to a scientific and public audience, he found fertile ground. It was not just that evolutionary ideas were in the air, though they were, but that there was a significant literature on historical and psychological development from primitive to civilized ways of life.

Enlightenment enthusiasm for universal histories of mankind, romantic concern with the special character of a people's history, linguistic and historical scholarship and geological research all shared a preoccupation with human origins. The Judaeo-Christian origin myth of the creation and fall of man had lost power in its biblical form; increasingly, it retained influence only in communities of believers isolated from the mainstream of public life. People turned for authoritative knowledge to the historical disciplines, which, scholars claimed, created not myth but science about origins, and evolutionary theory was in this tradition. The perception of time had transformed in the early nineteenth century. Educated opinion learned to imagine a time scale of millions of years for the Earth – 'deep time' – and hundreds of thousands, and perhaps millions, of years for humankind itself. Adam and Eve became mythic. Historical techniques for examining evidence were applied to the Bible, resulting in perhaps the most serious of all intellectual crises for the Christian faith.

In the late 1850s, at the same time as Darwin published, English writers, fascinated by archaeological evidence, began to refer to 'prehistory'. The term was a useful handle with which to grasp the new conception of human ancestry. Caves in France and England revealed human artefacts alongside long extinct animals, and collectors in river gravel paid workmen to dig out fashioned tools of great antiquity. Evidence for early people was found in Europe, and this required Europeans to accept the savage, primitive 'other' as the ancestor of European people, not just to think of 'the other' in terms of faraway exotic tribes. It became common belief that European people, if now civilized, are by ancestry connected to a primitive nature and a primitive mind. Indeed, the word 'primitive' came into common use. Imagination melded a perception of the living savage described by European travellers with the perception of early European people described by archaeologists. The result was a picture of a historically real primitive human, a savage ancestor still present within the psyche. The American psychologist G. Stanley Hall wrote in 1904: 'Most savages in most respects are children, or, because of sexual maturity, more properly, adolescents of adult size.' This linked savages, primitiveness, childhood and sexuality through the evolutionary laws of nature. Certainly, it threw no light on how such views themselves expressed social and moral categories. The opponents of women's emancipation drew on the same language and, at this time, linked femaleness to the primitive dimensions of human nature.

Also in the 1850s, excavations in a limestone cave in the Neander Valley near Düsseldorf, Germany, and in river gravels of the Meuse in Belgium, yielded human skeletal remains of great age. These remnants, only later clearly distinguished as Neanderthal man and the more recent Engis man, were the first fossil human finds. They did not, in the light of Darwin's work, provide evidence of a 'missing link', that is, they were not the direct ancestor of *Homo sapiens*. There was no known candidate for this position until the 1890s, when an explorer described so-called Java man (*Pithecanthropus*). (Scientists subsequently also denied that this was the 'missing link'.) In one respect relevant to psychology, Neanderthal man was striking. The cranial capacity, and hence the brain size, was more or less comparable with that of a modern human being. This was puzzling as most observers then thought of cranial capacity as a measure of mental capacity, yet the evidence from Neanderthal man suggested that the earliest known human type had as much capacity as a modern professor. The puzzle deepened when, in 1868, remains of an early truly human type were found in the Dordogne in France, and the leading French anatomists, Paul Broca and Armand de Quatrafages, deduced that a tall people with fine features had lived there in prehistoric times. As Thomas Henry Huxley (1825–1895) noted in *Evidences as to Man's Place in Nature* (1863), his widely read summary of the anatomical evidence linking monkeys and men, ancient and modern, the roots of humankind were much older than anyone had previously thought.

Then, also in the 1850s and independently of Darwin, Herbert Spencer (1820–1903) began the huge 'Synthetic Philosophy' that was to become the largest-scale evolutionary world view of the nineteenth century, with readers, literally, around the globe. Spencer's writing had great significance for psychology and, at times, more influence than Darwin's. However, because of the richness of the biological arguments and the appeal of the family gentleman, it is now Darwin's which is the household name.

Darwin and Spencer both went to great lengths to control their domestic circumstances, so that they could work, and both were often neurotically unwell. Spencer, though, was an irascible bachelor, while Darwin was a much-loved paterfamilias. Their paths barely crossed, and they divided the labour of philosophy and sociology and of scientific natural history between them. Spencer acknowledged that his early evolutionary thought ignored Darwin's central mechanism, natural selection, and he continued always to give priority to the mechanism of the inheritance of acquired characteristics. Darwin, for his part, was courteous in print. In private, however, he was noncommittal: he regarded Spencer as too abstract to be a good scientist. Darwin certainly was a natural scientist in a way Spencer was not, in the sense that he spent a lifetime in the detailed study of the history of plants and animals, and he was driven by a desire to explain detail. Yet Darwin constantly organized his information in relation to the *unifying theory* of evolution, as did Spencer, understanding 'that all observation must be for or against some view if it is to be of any service'. Darwin's work not only collated the observed facts into the general claim that evolution has occurred, but explored in depth a causal mechanism, natural selection, which explained it in terms of the material laws of nature. He set up what became, though not immediately, a comprehensive framework for biological research.

Darwin was happiest when at work on some detailed part of the living world, as in his studies of orchids or earthworms. Nevertheless, from at least the time of his world voyage, he was also fascinated by the question of man's place in nature. This interest culminated in two major studies in comparative psychology, *The Descent of Man, and Selection in Relation to Sex* (1871) and *The Expression of the Emotions in Man and Animals* (1872), drawing out the implications of his views on evolution for human nature. These books were widely discussed. The Victorian public had already been exposed to the idea of human evolution by the time he published, and this dampened the fire of criticism: his books appeared as contributions to a debate rather than as a shocking novelty.

Mankind was always at the heart of Darwin's experience of nature. On board the *Beagle* with Darwin were three Fuegians – Fuegia Basket, York Minster and Jeremy Button, indigenous people of the remote southern tip of South America, exposed to Western civilization in England, to be returned to their native land. Years later, Darwin described his experiences at Tierra del Fuego: 'The astonishment which I felt on first seeing a party of Fuegians on a wild and broken shore will never be forgotten by me, for the reflection at once rushed into my mind – such were our ancestors. These men were absolutely naked and bedaubed with paint, their long hair was tangled, their mouths frothed with excitement, and their expression was wild, startled, and distrustful.' This was a vivid image of the primitive. When he returned to Christian England, Darwin opened a series of notebooks, including those known to historians as the 'Metaphysical Notebooks'. Here, for his private use, he tried out what it was like to think as a materialist and to take a fully naturalistic approach to human nature. It did not cause him emotional problems to accept that humanity is the result of causal physical laws at work over time.

Darwin kept his eyes open for information relevant to human evolution, for example, on instinct. When he finally went into print in the *Origin*, he was at great pains to argue that natural selection could explain instincts since they were popularly believed to exemplify God's design in nature. Elsewhere in the *Origin*, he remarked, in a now famous understatement, that 'psychology will be based on a new foundation, that of the necessary acquirement of each mental power and capacity by gradation. Light will be thrown on the origin of man and his history.' Then, after a decade of debate about evolution, Darwin began the arduous task of turning his notes and thoughts on human beings into a book. Widespread acceptance of the evolution of animals and plants encouraged him to think that a public statement of his position on human nature would not prejudice sympathy for the cause of evolutionary theory in general. He also wanted to follow up in detail two topics of great interest to himself: sexual selection (preferential mate selection as a cause of selective reproduction and hence evolution) and the expression of the emotions.

Darwin found it difficult to shape his argument, and the *Descent* never achieved the intensity or command over its materials that made the *Origin* so persuasive a book. His argument was necessarily indirect since there was no record of human evolution. He maintained that if *Homo sapiens* differs, for example, in intelligence, from animals in degree but not in kind, then it is plausible to believe in human evolution. His strategy was therefore systematically to compare animal and human bodily and mental capacities in order to show that human beings have no capacities not shared, at least to a degree, with animals. As a result, he read into animal nature what is characteristic of human nature, and he used what he found in animal nature to confirm continuity between humans and animals. His opponents, by contrast, *defined* what it is to be human in terms of such characteristics as the moral soul, and thus Darwin's argument was intrinsically unpersuasive to them.

The *Descent* began with a comparison between human and animal bodies, and it sent readers away to check whether friends and relatives had points on their ears. This was familiar and uncontroversial ground; for example, in the 1860s there had been excitement over what, for the West, was the newly discovered gorilla. In two subsequent chapters, Darwin compared the mental faculties of animals and humans. Like Spencer, Darwin argued that there are inherited capacities or faculties, and he set out to show how these faculties – including those for language, the moral sense and intelligence – are present, in however an elementary form, in animals. Reason and language were crucial to Darwin's critics. In his approach to the evolution of these faculties, Darwin relied on earlier analyses (going back to Locke) which had traced the mind's content to experience. He traced intelligence to learning through sensation and language to expressive cries about what is felt. This did little to address deeper questions which plagued the analysis of reason, consciousness and linguistic meaning (or semantics). What it did do, though, as Spencer's work also did, was point to the comparative study of animals and the developmental study of children as the route to a science of psychology. The evolutionary outlook brought these topics to the centre of the stage, whereas they had been marginal interests earlier and sometimes even beneath the dignity of learned men concerned with the highest forms of reason. Darwin himself published observations of his eldest son, William, as a baby. It was a significant shift of direction in psychological interests. The moral sense received special and separate treatment; Darwin, after all, was a Victorian. With false modesty, he disclaimed ability to deal with such 'deep' questions but then boldly discussed the topic because 'no one has approached it exclusively from the side of natural history'. This approach was precisely what his opponents would not allow, since, in one step, it collapsed the distinction between mind and nature.

Darwin's step was a major contribution to the search for natural, psychological causes, as opposed to a philosophical or religious imperative, for moral action. He explained the existence of the moral sense, the Victorian conscience, by the imagined evolution of a social animal which also acquired high intelligence. Social instincts have evolved, Darwin thought, because the chances of an animal reproducing its kind increase if it co-operates with animals like itself in the search for food or in self-protection. He supposed that herd or family instincts increased the chances of survival of early human individuals who were physically weak by themselves. These instincts then began, he suggested, to be accompanied by reflective intelligence, and this caused the comparison of past actions with present outcomes, comparing, for example, impetuous selfish conduct with long-term suffering. This comparison, accompanied by pain, was the basis of conscience. Intelligence meanwhile, having given rise to language and the shared culture which language made possible, created the social world in which conscience was reinforced by custom embodied in public opinion. Lastly, Darwin believed, the inheritance of acquired patterns of reflective activity from one generation to the next enlarged moral capacities into a fully civilized moral sense. Thus we can see how altruistic activity may have evolved by natural selection. He thought a process which favours 'survival of the fittest' had in fact favoured people who contributed most to society's welfare – and thus contributed also to their own. It was an argument which attempted to reconcile 'the golden rule', do as you would be done by, with an account of the *usefulness* of moral actions.

Argument about Darwin's account of social evolution, and about the very possibility of a naturalistic basis for ethics, is still with us. Even those who took the label 'Darwinist' in the nineteenth century reached different conclusions about what moral instinct people inherit and how they should act. Darwin's half-cousin Francis Galton, for example, believed that people have very unequal innate capacities. In his view, therefore, social and moral progress requires political conditions in which individuals with innately superior capacities are more highly valued than other people, especially when it comes to the production of children. The utopian anarchist, the Russian prince P. A. Kropotkin, by contrast, argued that people, like animals, have a social instinct and are naturally altruistic; progress therefore lies with the removal of present institutions to allow the spontaneous organization of communal life.

Darwin said little about language and culture when he discussed the origins of human morality. He did not see the evolution of language and culture as a revolution in the evolutionary process, as Alfred Russel Wallace (1823–1913), his co-theorist on natural selection, suggested. Darwin thought that natural selection continued to operate in modern society, though modified by custom, law, morality and religion. Wallace, in contrast, in the 1860s, raised but barely followed up the insight that the advent of rational consciousness transferred the motor of evolutionary change from nature to conscious human action: 'A being had arisen who was no longer necessarily subject to change with the changing universe – a being who was in some degree superior to nature, inasmuch, as he knew how to control and regulate her action.' He implied, once mind and culture had evolved, that evolution became a *human act*. It followed that there may be new forms of change, psychological processes and modes of sociality not found in nature. Darwin never assimilated this perspective and continued largely to think about human evolution in terms of the selection of individual characteristics and not in terms of the history of societies in which individual characteristics may be unimportant. Here is one of the sources of the recurrent disagreement about the relationship between culture and nature in human affairs.

There were two further significant topics in Darwin's natural history of human beings. In the *Descent*, he hoped to clear up the vexed question of the origin of human races and racial differences. Evolutionary theory, Darwin believed, put the last nail in the coffin of the ancient belief that primitive people had degenerated from a higher condition. Along with Wallace, he thought it made redundant the debate between monogenists and polygenists about whether races have a single origin or separate origins: from the evolutionary perspective, it was a matter of how far back one looked. Darwin had high hopes that sexual selection would explain the origin of racial differences, as well as secondary sexual characteristics, such as features of hair or physiognomy which have no apparent relevance to survival. If sexual selection explains the colouration of butterflies, it may explain human colouration too. He also used the theory of sexual selection to suggest how human activity which does not appear useful, in dance or the appreciation of beauty, for instance, had evolved. Darwin filled more than half his book with the evidence for sexual selection among animals – the sort of material with which he was much more at home than mental science.

Darwin was equally fascinated by a second topic, the expression of the emotions – the study of which, he thought, was a rhetorically powerful but empirically precise way to compare people and animals. His argument had bite. He traced his interest in the subject to the work of the anatomist Charles Bell (1774–1842), whose beautifully engraved plates explored 'his view, that man had been created with certain muscles specially adapted for the expression of his feelings' – that God designed the face as the outward expression for the soul within. It was a delighted Darwin who compared animal and human expressions – with pictures of a snarling dog and a furious child – and suggested that natural selection rather than God's purpose explains the similarities.

*The Expression of the Emotions* was an extended study of physiognomy (the practical psychological art of reading human character) that had fascinated earlier generations, extended to the whole body. In Darwin's hands, this art became part of physiology – an attempt to explain expression. For instance, tears, Darwin thought, are a habit acquired by the species from the contraction of muscles around the eyes and the engorgement of the eyeballs by blood during pain, and pain, or the thought of pain, now leads to tears. His book collected together descriptions of animal expression and a mass of observations of children, savages, lunatics, actors and everyday people to build up a picture of the emotional human animal and its innate repertoire of expressions. He discussed anxiety, despair, joy, love, devotion, sulkiness, anger, disgust, surprise, fear, modesty, blushing and much else besides, drawing ordinary psychological language into the orbit of physiological explanation.

Biologists now much admire this study as a pioneering work in ethology (the study of animal behaviour), and as an anticipation of psychological research which observes human life as the exhibition of behaviour rather than mental purposes. It was a fitting capstone for Darwin's natural history, in which photographs of manic lunatics, a diary of his young son, the coloured rump of the mandrill monkey in sexual display, the killing of indigenous people, male nipples and even the nagging Victorian conscience were grist to his mill.

Thus, the evolutionary narrative encouraged a turn from the analysis of reason to examine emotions, so-called primitives, children, instincts and animals, and it directed attention to what people do rather than to what they think. It directed attention to developmental processes rather than fixed states. Together with the new emphasis on mind as the activity of brain, it made psychology, if still the study of mind, the study of mind understood as functions, activities in which people deal with the world, rather than the study of reasoning or spiritual life. Of course, all this had beginnings before Darwin – some historians would go back to Aristotle – but it was in the late nineteenth century that the change took deep hold.

Spencer was the most systematic, if at times tedious, theorist of the new intellectual outlook. Philosophers as far back as Plato had likened human communities to organisms and drawn attention to the order, dependencies of parts and integration of civilized society and animals alike. Henri de Saint-Simon, a French social theorist in the early nineteenth century, Comte and then Spencer filled out the comparison with new ideas about what actually gives rise to living organizations. Spencer reconceived organization as adaptation produced over time, and he argued that evolution explains even the most complex integrated system in terms of natural laws. From this perspective, to understand a complex system, like a mind, the social world or a commercial business, is to show how its parts have evolved by interaction with each other in ways which have sustained the life of the whole. Psychological (and sociological) thought at the end of the nineteenth century substantially adopted this perspective, followed Spencer's lead and, inspired by the theory of evolution, attempted to understand individual and social life through the way parts serve organized wholes. This was the basic orientation of functionalism.

Spencer had the virtues and weaknesses of his upbringing in the provincial English world of Nonconformist Protestantism. An independent-minded individualist, he made his own way from railway engineer to sage, opposing established religion but promoting the morality associated with it. As a young man, he was enthusiastic about phrenology; a reading of his head found bumps for firmness, self-esteem and conscientiousness, and the conclusion was that 'such a head as this ought to be in the Church.' Like Comte before him, he constructed a philosophical synthesis with the progress of knowledge in natural science as the main theme of human history. In doing this, he took up the biological language which describes organisms in terms of structure and function, and he applied this language to understanding psychological life and human society. To understand the mind, Spencer argued, we must understand what the mental functions of the brain enable us to do. Psychology cannot develop as the analysis of *mind*; it must be the analysis of *active people*.

Spencer, unlike Darwin, contributed no new observations, but he did provide a conceptual framework for psychology (and sociology) as a subject comparable with the natural sciences. The framework involved two major principles: continuity of natural law and utility. He embodied the former in his description of the evolutionary law of directional change at all levels of reality, from the cosmos to the free market. At each level, he argued, by applying the principle of utility, we can understand the laws at work in terms of adaptive integration to conditions. These principles were very abstract – and Spencer's style enhanced this impression: 'Evolution is an integration of matter and concomitant dissipation of motion; during which the matter passes from an indefinite, incoherent homogeneity to a definite, coherent heterogeneity.' Nevertheless, his tenacity – due to his being 'the first to see in evolution an absolutely universal principle', as William James wrote – fostered a seachange. The notion that adaptation occurs through experience, and that adjustment of an internal state under the impact of an external state produces a stable condition, became Spencer's formula for the description of any system whatever. He generalized evolutionary theory. In the mid-twentieth century, for instance, this provided the conceptual basis for the approach to decision-making called systems analysis.

Spencer developed his evolutionary thinking independently of Darwin and first applied it in *The Principles of Psychology* (1855). Whereas Locke and his followers had treated experience as the way an individual mind acquires knowledge and thereby responds constructively to the external world, Spencer treated experience as a process over extended time: the way one generation builds on the experience of ancestors and hence the way animals and human beings evolve and adapt. This is what he meant by the continuous adjustment of inner to outer relations. In Spencer's hands, empiricist psychology therefore became the study of mind and culture as products of adaptive integration to previous conditions, a process which, Spencer believed, inexorably results in human progress. Our psychology is the history of the human race.

Earlier writers who derived the content of mind from experience, like Condillac or Hartley, awkwardly supposed that each individual mind builds up its mental content from scratch. By contrast, Spencer – though he too derived the content of mind from experience – emphasized that each individual inherits mental functions from previous generations. This was an important step in two ways. First, common opinion had always found implausible Locke's image of the newborn mind as a blank sheet, since babies, like animals, appear to be born with emotions and instincts. Further, as J. S. Mill argued against his father, James Mill, it is simply incredible to reduce the emotional life to a calculus of pleasures and pains. Darwin made much the same point and argued that a theory of the emotions requires a theory of inherited instincts. Evolution, Spencer believed, embeds experience in the inheritable structure of the nervous system. A mental event, over time, changes from a mental process to a nervous structure. This explains, for example, how 'the most powerful of all passions – the amatory passion – [is] one which, when it first occurs, is absolutely antecedent to all relative experience whatever.'

There was more to this than the simple expansion of the psychology of experience to include instincts and emotions. They had long been important evidence in the arguments of idealist and Christian opponents of empiricist psychology, and British moralists had lovingly described animal instincts, like bees building a hive, and human capacities, like the moral sense, to illustrate God's design in nature and human nature. Conservative academic philosophers argued that the mind possesses categories which, of logical necessity, cannot originate with experience and are God-given. Spencer believed he was able to pull the rug out from under these Christian writers who he thought bulwarks of the conservative political order in Britain. Evolution, he argued, demonstrates that emotional and rational capacities are indeed innate but that they nevertheless derive from experience: evolutionary experience. Thus, what had appeared to be convincing evidence for Christian idealism reappeared as evidence for a natural science of psychology.

Further, Spencer thought he had put to rest the philosophical dispute between idealists (like Leibniz or Kant) and empiricists (like Locke or the Mills) about the origins of knowledge. He accepted that the individual mind shapes knowledge with a priori categories but held that these categories are in fact a posteriori in an evolutionary sense. This psychological answer to a philosophical problem was exactly the kind of argument that philosophers, led by G. Frege in the 1890s, reacted against when they laid the basis of what became twentieth-century analytic philosophy. Spencer, however, intended to carry naturalistic or scientific ways of thought into philosophy, and, in late twentieth-century Anglo-American philosophy of mind, there was renewed sympathy for this programme.

It took a while for people to appreciate the significance of these arguments. Spencer's early work on psychology went largely unread, though at the end of the 1850s he persuaded enough people to back him financially to write a 'synthetic philosophy'. It began with *First Principles* (which appeared serially 1860–62) and concluded 30 years later with studies on sociology and ethics. The second edition of the work on psychology (1870–72), in which Spencer reshaped his argument, did attract attention. Thereafter, Spencer moved towards the ultimate goal of his endeavours, the foundation of ethics and an individualist politics grounded on the natural law of progress, rather than towards the detail of psychological research.

## TOWARDS A PSYCHOLOGY OF WHAT PEOPLE DO

Darwin, Spencer and their generation attempted speculative reconstructions of the evolution of mind. Later psychologists, more preoccupied by methodology, ruled this out and instead opted to study the function of parts in wholes as directly observable in the present. This, in practice, meant the observation of people or animals doing things in laboratories, and by this route evolutionary functionalism became of overriding importance in the United States, where, in the last decade of the nineteenth century, psychology first developed as a large-scale academic discipline. Functionalist analysis also became standard practice in the social and management sciences, which began to develop about 1900. Psychologists sought to learn about the adaptive integration of people in institutions and in society at large, and they rarely asked questions about whether people have a common political goal or what form political society should have.

Though Bain never fully assimilated an evolutionary outlook, he also contributed to this way of thought focused on acting and adaptation. When he brought association psychology into relation with nervous physiology, he took up the idea – already in the literature – that experience is not a passive affair but the result of a person's (or animal's) spontaneous activity: action precedes sensation. This addressed the standard criticism of empiricist psychology that it denigrated mental life by rendering it as a mere passive receptivity to sensations. Bain's belief about activity opened a view of learning as a kind of trial and error, in which a person first acts and then experiences pleasurable or painful consequences, which leads to adjustment. This was to be a standard claim in twentieth-century learning theory in the U.S., discussed later, dubbed 'the law of effect' by E. L. Thorndike and central to B. F. Skinner's science of operant conditioning.

In Europe, the reception of evolutionary ideas was greatly complicated by different beliefs, a number of which predated Darwin and Spencer, about what evolution actually meant. Spencer was, if anything, more influential, at least in psychology, than Darwin. When Théodule Ribot (1839–1916), the leading exponent of reform in psychology in France in the 1870s, set about displacing established *psychologie* in France, psychology based on belief in the spiritual unity of the mental self, he turned to both German experimentalists and English physiological psychologists. Spencer had pride of place in *La Psychologie anglaise contemporaine* (1870; *Contemporary English Psychology*). Ribot self-consciously referred to 'the new psychology', and many other writers used the same rhetoric of a new stage in objective knowledge. According to Ribot, writing in 1879: 'The new psychology differs from the old in its spirit: it is not metaphysical; in its end: it studies only phenomena; in its procedure: it borrows as much as possible from biology.' In practice, there were many varieties of new psychology, and the phrase has misled historians into thinking in terms of some kind of single shift. It did, however, clearly denote the experience of something old passing, the passing of a way of thought looking to metaphysics and religion rather than empirical science for truth.

Darwin, George John Romanes (1848–1894, to whom Darwin entrusted notes on instinct) and, in the 1890s, C. Lloyd Morgan (1852–1936), all worked systematically on animal instinct and intelligence. They were aware that their views tended towards anthropomorphism, the projection of human attributes into where they do not belong. The obvious problem, familiar to everyone, is knowing how far down the animal scale to attribute consciousness: to the pet dog, yes, but to her fleas? Darwin and Romanes were so keen to demonstrate evolutionary continuity that they did not find it much of a problem; indeed, they wanted to show animals are like humans, and vice versa. But Morgan and later observers thought it compromised objectivity. It therefore became necessary, in studying animal psychology, to devise methods for objective description rather than projecting the experience of the human mind. This, once again, encouraged the study of behaviour rather than the study of mind. Morgan proposed what was to become cited as a principle: in no case is 'an animal activity to be interpreted as the outcome of the exercise of a higher psychical faculty, if it can be fairly interpreted as the outcome of one which stands lower on the psychological scale'. This was a methodological rule, from a non-scientist's point of view, at the expense of the interest animals have in people's lives. It was indicative of a pattern which became common as scientific psychologists set out to demarcate what they did, objectively, from what ordinary people did, subjectively, in understanding psychological life.

There was a public audience for varieties of new psychology, especially on the animal–human comparison, just as there was for evolutionary ideas (themselves often expressed in cartoons making monkeys men, and men monkeys). Spencer (on sociology), Bain (on mental science) and Romanes (on animal intelligence) all published volumes in the International Scientific Series, which was a major venture in joint North American and British publishing in the public understanding of science. Another ardent Darwinian, John Lubbock, who was also a writer on prehistory, published in the Series on ants, bees and wasps, and wrote much about social instincts. A further evolutionary thinker and psychologist who gained a public audience, and who moreover retained that audience, was William James (1842–1910). For some, he is America's greatest philosopher; for many, he is the greatest writer in English among psychologists, the writer whose idiosyncratic vividness of expression sustains hope that it does not have to be a boring discipline. His sister, Alice, kept a now famous diary. William's younger brother Henry wrote novels both profoundly admired and scorned by some for the extreme refinement and the exquisiteness of descriptions of subjective psychological experience. But William the psychologist has presented historians with a puzzle, especially historians who have portrayed the emergence of different so-called schools of psychology in the twentieth century, since James left no school and psychologists found much to criticize in his work.

The greatness of James, I think, is first in the writing, second in the integrity with which he balanced ways of thought – all of which appeared necessary even if they were mutually contradictory – and third, in the argument for the pragmatic theory of truth. His intellectual and emotional life was a struggle to accommodate the facts of physiological science and spiritual ideals in an evolutionary outlook. It had a sharp personal edge. He had a need to assert his will as an individual in a world in which, scientists were concluding, events happen by necessity.

After an unusually diverse education, much of it in Europe, and which included qualifying in medicine, from 1875 James taught a course in psychology at Harvard. He used Spencer's *Principles of Psychology* as his text. James drew together the threads of psychological experimentation current in Germany, physiology, medicine and functionalist forms of explanation informed by the theory of evolution. Aided by a colourful style and vivid use of metaphor, he produced what for some psychologists remains the masterpiece of their subject: *The Principles of Psychology* (1890). Thereafter, James was more preoccupied by philosophical matters, and he turned a search to understand the mind in terms of its functions in the evolutionary process into a theory of knowledge called pragmatism. This is a theory which judges the truth of a proposition in terms of its consequences for action. His sympathy with experience as a form of truth if it acts as a guide to life – even if the experience does not appear to be scientific – led him to studies of psychical phenomena (he worked with the Boston medium, Mrs Piper), from which many other psychologists wanted to distance themselves. The development of experimental methods during the 1890s was, in a number of settings, part of a self-conscious attempt by academic psychologists to distance themselves from the public perception of psychology as the study of psychical phenomena. James, however, valued the connection and looked to it to reveal the full range of human experience. This same sympathy led James to a major study in the psychology of religion, *The Varieties of Religious Experience: A Study in Human Nature* (1902).

James contributed in the 1870s to a lively British debate about what evolutionary continuity means for the relation of mind to brain. His position at this time contained the seeds of much of his later philosophy. He reasoned that if conscious mental activity is the outcome of the evolutionary process, then, like a bird's plumage or any other aspect of an animal or plant, it must have adaptive function. It could not be correct, therefore, to treat consciousness as an epiphenomenon, as if it were merely like the whistle on a steam engine, adding nothing to the power of the engine. This argument also expressed James's deep personal need to believe that his choices, indeed his life, made a difference. He therefore opposed Huxley who, carried away by his own metaphor, likened consciousness to the sound of a bell: 'The soul stands related to the body as the bell of a clock to the works, and consciousness answers to the sound which the bell gives out when it is struck.' Instead, James supposed that consciousness has consequences, and, at one stage, he described 'conscious interests', a purposeful dimension to our conscious life that makes a difference. This was the basis of pragmatism: 'Mental interests, hypotheses, postulates, so far as they are bases for human action – action which to a great extent transforms the world – help to make the truth they declare. In other words, there belongs to mind, from its birth upward, a spontaneity, a vote.' What consciousness does is attend selectively.

Thinking along these lines, James took seriously human values and purposes as real forces in the world and thus, so to speak, he reinserted values and purposes into nature from where they were being removed by mechanistic science. If he did not achieve a synthesis which satisfied philosophers, he encouraged himself and others, though fully persuaded by the power of natural science, to think that science does not destroy belief in conscious agency. Rather, James, taking the evolutionary perspective, pointed towards belief in science as the adaptive means by which humanity, conscious of its own power, directs the cosmos towards its own well-being. This was the intellectual's version of the American dream: everything is possible, everything depends on human action. James was not the only scientist to argue that human consciousness makes a difference to the evolutionary process. I noted earlier that Wallace made this point. Two other psychologists in the 1890s also took up the theme: Morgan in England and James Mark Baldwin (1861–1934) in the United States. Baldwin (with Cattell) founded the *Psychological Review* and headed psychology laboratories in Toronto, Princeton and then Baltimore.

The writings of another American philosopher, John Dewey (1859–1952), also profoundly imbued with the evolutionary outlook, later stood out for the quality of theoretical argument, though they barely touched day-to-day psychological research at the time. Dewey's paper, 'The Reflex Arc Concept in Psychology' (1896), re-described the central unit of the mechanist analysis of the nervous system – the reflex – as a process. He argued that the structural elements of the reflex (sensation, central connection, motion) are not distinct; a proper understanding of the reflex lies with the description of its contribution to the adaptation of the organism *as a whole*. Similarly, he thought, psychologists should not study mental elements independently of the integrated activity of which they are part – the action that gives rise to them and the action to which they contribute. Critics of behaviourism in the 1920s made the same point and berated psychologists for treating life as if it consisted of stimuli causing responses rather than a continuous process of adaptive transformation.

It was no coincidence that Dewey was familiar with Hegelian philosophy, with its approach to mind as a historically unfolding process, as well as the British theory of evolution. Dewey, like James, was a man with high social ideals: he looked to his academic occupation to provide leadership as society evolved and adapted, as he trusted it would, into an enlightened democracy. Associated with the work of G. H. Mead, who shared the same background in German philosophy, Dewey's ideas became a significant influence on social psychology and social theory as well as on the philosophy of pragmatism.

To conclude: between 1830 and 1900, the status of evolutionary theory changed from being a glint in the eye of radical natural historians to become the *raison d'être* of progressive democratic society. A debate which was at one level about the empirical evidence for the origin of species by descent with modification was, at another level, about the terms in which to understand what it means to be human. Similarly, the debate about the physical embodiment of mind was at one level about the empirical evidence for the dependence of mind on brain, while at another level it was about what sort of psychology would do justice to the full range of human experience. The biological outlook, both physiological and evolutionary, fostered scientific explanation in terms of functions, and this underwrote in intellectual terms the liberal political commitment to act to adapt humanity and make possible a better life. Psychologists (and social scientists) did not just come to believe in evolution but came to believe in their science as the highest stage of evolution. It is therefore time to match this discussion of ideas with an account of how psychology did, in fact, become a discipline, a social institution, even if it did not achieve unity of outlook, in the half-century before 1914.

## READING

This chapter discusses developments in intellectual history and the history of science that historians have covered in enormous detail. On nineteenth-century thought, see M. Mandelbaum, *History, Man, & Reason: A Study in Nineteenth-Century Thought* (Baltimore, ML, 1971); O. Chadwick, *The Secularization of the European Mind in the Nineteenth Century* (Cambridge,

1975); J. W. Burrow, *The Crisis of Reason: European Thought, 1848–1914* (New Haven, CT, 2000). For scientific 'materialism' (a very loose word), see: F. Gregory, *Scientific Materialism in Nineteenth Century Germany* (Dordrecht, 1977); A. Kelly, *Descent of Darwin: The Popularization of Darwinism in Germany, 1860–1914* (Chapel Hill, NC , 1981); G. Dawson, *Darwin, Literature and Victorian Respectability* (Cambridge, 2007); and D. Joravsky, *Russian Psychology: A Critical History* (Oxford, 1989) – a fascinating study, though the author was denied access to Russian archival sources and Russian psychologists disliked the story he tells as he finds greater understanding in literature.

For phrenology, R. M. Young, *Mind, Brain, and Adaptation in the Nineteenth Century: Cerebral Localization and Its Biological Context from Gall to Ferrier* (Oxford, 1990), which also discusses Bain and Spencer; and for hypnotism as a public enthusiasm in Britain, A. Winter, *Mesmerized: Powers of Mind in Victorian Britain* (Chicago, 1998). For the beginnings of nervous physiology, E. Clarke and L. S. Jacyna, *Nineteenth-Century Origins of Neuroscientific Concepts* (Berkeley, CA, 1987). On physiological psychology: K. Danziger, 'Mid-Nineteenth-Century British Psycho-Physiology: A Neglected Chapter in the History of Psychology', in *The Problematic Science: Psychology in Nineteenth-Century Thought*, ed. W. R. Woodward and M. G. Ash (New York, 1982), pp. 119–46; R. Smith, 'The Physiology of the Will: Mind, Body, and Psychology in the Periodical Literature, 1855–1875', in *Science Serialized: Representations of the Sciences in Nineteenth-Century Periodicals*, ed. G. Cantor and S. Shuttleworth (Cambridge, MA, 2004), pp. 81–110; J. Bourne Taylor and S. Shuttleworth, eds, *Embodied Selves: An Anthology of Psychological Texts, 1830–1890* (Oxford, 1998). There is also a large and relevant literature on nineteenth-century psychiatry; overall, see E. Shorter, *A History of Psychiatry: From the Era of the Asylum to the Age of Prozac* (New York, 1996).

For evolutionary thought: J. R. Durant, 'The Ascent of Nature and Darwin's Descent of Man', in *The Darwinian Heritage*, ed. D. Kohn (Princeton, NJ, 1985), pp. 283–306; R. M. Young, *Darwin's Metaphor: Nature's Place in Victorian Culture* (Cambridge, 1985); R. J. Richards, *Darwin and the Emergence of Evolutionary Theories of Mind and Behavior* (Chicago, 1987). J. H. Brooke, *Science and Religion: Some Historical Perspectives* (Cambridge, 1991), spread knowledge of multiple relations between science (including evolution) and religion. For Darwin, there is a fine biography: J. Browne, *Charles Darwin, Vol. 1: Voyaging* and *Charles Darwin, Vol 2: The Power of Place* (Princeton, NJ, 1998, 2005). Darwin's *Descent of Man* has a useful introduction by A. Desmond and J. Moore (London, 2004). R. Boakes,

*From Darwin to Behaviourism: Psychology and the Mind of Animals* (Cambridge, 1984) clearly describes animal psychology. James appeals to biographers; for an overview, P. Croce, 'Reaching beyond Uncle William: A Century of William James in Theory and in Life', *History of Psychology*, XIII/4 (2010), pp. 351–77.

CHAPTER THREE

# Shaping Psychology

> My experience is what I agree to attend to. Only those items which I notice shape my mind – without selective interest, experience is an utter chaos. Interest alone gives accent and emphasis, light and shade, background and foreground – intelligible perspective, in a word.
> William James

## NATIONAL PSYCHOLOGIES IN THE LATE NINETEENTH CENTURY

Debate about education invigorated public life throughout the nineteenth century. Reformers looked enviously at Germany, where academic culture was especially esteemed and where learning, organized along disciplinary lines, successfully created new knowledge and trained new scholars. Germany appeared to demonstrate that support for specialized sciences was essential to the modern nation state. Humiliated by war with Prussia in 1870, the French Third Republic assimilated lessons from the German system of higher education. Italy employed German, or German or Austrian-trained, scientists in its universities in a drive to construct a modern nation, before and after unification in 1870. Idealistic young Russians came to Germany, Austria and Switzerland in search of higher education, and they returned with Western ideas. English speakers also travelled to continental Europe to get advanced, specialized training.

Eighteenth-century professors were, by and large, generalists, with an audience of male students who expected to go into public life, and if they published at all, it was for a small non-academic readership. It was a significant innovation when the new and reformed universities, led by Göttingen and Berlin (founded 1809–10), restructured teaching and research along disciplinary lines under the aegis of the philosophy faculty. The faculty acquired status equal to or greater than the senior faculties of law, medicine and theology, to which it had earlier been subservient. The new academics were disciplinary specialists who increasingly wrote primarily for their peers and taught a new generation of scholars in their subject. Disciplines like philosophy, history, philology, chemistry and physiology came into existence as distinct social entities. The audience for specialist research and the audience for general literature began to be distinct, and the notion of an amateur as someone who does not really know what they are talking about came into use. The new disciplines did not include psychology as a distinct entity. Scholars who addressed psychological topics were likely to be philosophers (for example M. W. Drobisch, a philosophical mathematician) or physiologists (such as E. H. Weber), and there was no clearly separate audience for what they wrote. When, later, psychology also developed as an academic discipline, the result was a tension, which persists, evident in all fields but sharp in psychology's case, between academic and public goals.

As the universities developed into strong and relatively independent institutions, they spread values different from those informing the day-to-day practical concerns of ordinary men and women and from those of people in commercial and political life. Academics took up *Bildung* (literally, 'formation') as the ideal value, denoting the achievement of wholeness by a person and a culture in pursuit of the good, the true and the beautiful. Spokesmen declared that only individuals ('cultured people') and the collective culture working in harmony could achieve this end. Thus, writers idealized Goethe and his age. The outcome for the sciences – and at this time all disciplines, including theology, insofar as they pursued rationally grounded knowledge, were known as sciences – was the pursuit of knowledge, pure science, as an end in itself. This ideal was always in tension with support for the universities from rulers, patrons or society at large. When psychologists pushed to form themselves into a discipline, they were, from the beginning, acutely conscious of this tension. Only in the twenty-first century was the tension resolved, to the satisfaction of politicians, with the straightforward demand that academic activity first justify itself in commercial terms.

The confrontation between university ideals and the public demand for practical knowledge is well illustrated in the development of higher education in the United States. The new country aspired to be a political community founded on rational principles, free of religious intolerance and responsive to the values of humanity, in a way the older European powers were not. American college education in the first half of the nineteenth century reflected little of the changes taking place in scholarship in Germany. Colleges aimed to fit young gentlemen into their place in the professions and public life. Teachers equipped students with moral and social notions of good character, and the teaching of psychology began in this context. One of the most eminent representatives of this tradition of moral philosophy teaching, a Scotsman, the Reverend James McCosh (1811–1894), who became president of the College of New Jersey (Princeton College) in 1868, taught a curriculum which included descriptive psychology as part of moral training.

School education was also a primary concern. At least since the early eighteenth century, progressive writers had linked a country's wealth to the state of education of its people. The high literacy rates in England and its commercial leadership appeared to prove the point. In the nineteenth century, pressure to introduce mass education accompanied the opportunities which industrial and urban life created, alongside spreading nationalism and republican and democratic sentiment. In France in the mid-century and then in Britain in 1870, legislation established the framework for universal primary schooling. The principal motive in France, where education for children between six and fourteen became compulsory in 1882, was political, and aimed at creating a sense of common national identity across a large and largely rural country. In Britain, where elementary education became compulsory in 1880, the intention was to enable people to exercise sound political choice and to become economically self-reliant. Mass education followed elsewhere too. This was of enormous importance for the field of psychology. The practical aims of education existed alongside and shaped the scientific aims of would-be specialist psychologists.

In 1800 there was no department of psychology and no occupation of psychologist. By 1903, in the United States there were at least 40 psychology laboratories; the subject awarded more doctorates than the other sciences, except for chemistry, physics and zoology; a professional society (the American Psychological Association (APA) had been founded in 1892); and there were specialist journals, headed by the *American Journal of Psychology* (1887) and the *Psychological Review* (1894). None of this had existed even in 1880, when only James at Harvard taught 'the new psychology' and used some laboratory facilities. Psychology thus grew remarkably rapidly as an academic discipline. But the case was unique. Even in Germany, from where much of the inspiration and training for the American discipline had come, the emergence of psychology as a separate field was not at all so clear-cut. The situation in different countries varied considerably, and this justifies reference to national histories and country-by-country discussion.

The universities in France, until late in the nineteenth century, were examining boards rather than centres of training and research. It was *les grandes écoles* and institutes which were home to science. Running through education at every level, as through society, was a continuing struggle between Catholic institutions and the interests of those committed to a secular republican state. In this context, two determinedly secular students of the École Normale Supérieure, Ribot and Hippolyte Taine (1828–1893), elaborated intellectual and institutional support for a non-Catholic, anti-Cousin approach to psychology. They opposed the *spiritualisme* which grounded psychology in belief in a soul existing unified and distinct prior to experience. For resources in their campaign, they turned to both British and German literature, the former for its empiricist philosophy and evolutionary outlook, the latter for its physiology and experimental methods. They also built on a native French tradition of clinical studies of the embodied morbid mind. This last tradition gave rise to an image of French psychology as peculiarly based on clinical methods.

Taine, later famous for his history of France, attacked French conservative philosophy with a book on British psychology, *De l'intelligence* (1870; *On Intelligence*), and his general style of writing was a call for a secular science of mind. Ribot, besides writing studies of British and German psychology and translating Spencer, in the 1880s published a series of accessible books on disorders of will, memory and personality, tying psychology closely to medicine and the authority of argument from the individual case. He promoted children, primitive people and the mad as the three ideal subjects of scientific psychology: these groups, he believed, throw into relief the normal, male, rational mind, and they are, therefore, natural psychological experiments. Ribot wrote: 'The morbid derangement of the organism . . . produces intellectual disorders; anomalies, monsters in the psychological order, are to us experiences prepared by nature, and all the more precious as the experimentation is more rare.'

The establishment of a chair in experimental and comparative psychology at the Collège de France, to which Ribot was appointed in 1888, signalled his success in establishing new psychology in France. Ribot was an eclectic writer and did not undertake original research, but his publications, editorship of the *Revue philosophique* and new position made him a pivotal figure in fostering psychology as a natural science. He advised Alfred Binet (1857–1911) early in the latter's career, and Binet went on to found, in 1895, the journal *L'Année psychologique*, to direct the psychological laboratory (founded in 1889) at the Sorbonne and to formulate tests which were the beginning of the measurement of intelligence. Ribot also associated socially and intellectually with the famous neurologist J.-M. Charcot (1825–1893), whose descriptions of hysterics, hypnotized subjects and religious ecstatics were part of a secular campaign against what these Paris intellectuals considered ignorance and religious superstition. This was a world with which the young Freud was intimately familiar.

A fascination with abnormal mental states drew many people to take an interest in psychology. Most dramatically, there was an overlap between investigations in psychology and in psychic phenomena, and the conflation of the two areas was so strong that the academic psychologist Binet struggled to get the public to see a difference. The psychologist and physician Pierre Janet (1859–1947), who followed Ribot in the chair of psychology among the academic elite at the Collège de France, remarked that the chair existed only because of cases of multiple personality. Janet meant that cases of multiple personality, which seized imagination at this time, proved the existence of something other than a unitary self, and this led to the conviction that the self had to be a subject for medical science rather than religion. Ribot wryly noted: 'It is but natural that the representatives of the old school, slightly bewildered at the new situation, should accuse the adherents of the new school of "filching their ego".' Janet built his reputation on extended researches, published as *L'Automatisme psychologique* (1889; *Psychological Automatism*), which he carried out in Le Havre on 'Léonie', a woman remarkably gifted when hypnotized and possessing a multiple personality. When the first international congress of psychology took place in Paris in 1889, Charcot was its president and one-third of the papers presented were on hypnotism. Subsequently, in reaction against the association between psychical research and psychology, psychologists called for a congress on experimental psychology. Nor were these circumstances unique to France: for example, the first society for experimental psychology founded in Berlin, in the 1880s, undertook psychical research.

The Netherlands and Scandinavian countries were much influenced by the German academic system, and scholars trained in Germany found opportunities to teach psychology as a new scientific speciality. The professor of philosophy in Groningen, Gerard Heymans (1857–1930), founded the first psychological laboratory in the Netherlands in 1892, though his appointment was in the history of philosophy, logic, metaphysics and the science of the soul. The science of the soul was taught elsewhere in the Netherlands, but Heymans stood out because he treated it as an empirical or experimental as well as a philosophical subject. Special chairs in psychology were then founded in the 1920s, but, unlike the United States, the system of education provided no basis for rapid expansion because pedagogues remained committed to a philosophical, moral and religious conception of their occupation. Psychology began to expand after the First World War, when church leaders turned to psychology for help in addressing the materialism and spiritual vacuum they saw around them in urban and industrial communities. Until the 1960s, Dutch society was divided – 'pillarized', as the Dutch say – between Protestant and Catholic communities and with separate universities. The communities nevertheless shared pastoral values and an interest in anti-materialist psychology and social science. After 1945, there was a spectacular growth in psychology and the social sciences generally. This was attributable in part, it would seem, to the Dutch public's readiness to accept expert guidance in life as a substitute for religious leadership. Alfred Lehmann (1858–1921), who visited Leipzig in order to familiarize himself with psycho-physical apparatus, brought the experimental methods of Wundt (shortly to be discussed) to the university of Copenhagen in 1886. The university's expectation was that he would relate his work to philosophical questions, but Lehmann preferred to focus on more narrowly defined psychological topics, such as recognition, a problem in perception which he developed by examining the association of sensory elements.

In those parts of Europe where liberal and educated people were a distinct minority, reformers pinned their hopes on a combination of national liberation or revival, expansion of free economic activity and modern scientific learning. The educated elites often studied in the German universities, acquired specialist knowledge of subjects like experimental psychology and then returned home to spread a scientific approach to social questions. Much of Europe in fact remained rural, remote and economically medieval even at the end of the nineteenth century. Students from this background faced conditions like those of people from around the world – from Turkey, China, Japan and Argentina, for instance – who, studying in Germany or Paris, acquired a Western outlook and, after returning home, worked for modernization. The young men and women who did study abroad, from towns like Helsinki (then Helsingfors, the administrative centre of a Russian province) or Bucharest, established new subjects like psychology to the limited degree that local circumstances permitted. Lecturing on experimental psychology or psychology in education, they pushed open windows to modern ways of thought. After 1918, with countries like Romania gaining political independence, there were some opportunities for institutional growth as a response to the practical needs of education and modern commercial operations.

In Italy, there were many connections between the educated elite, which formed the backbone of those set on modernizing the country, and the German and Austrian universities. Scholars like Vittorio Benussi (1878–1927), who was interested in perception, visual illusions and, later, psychoanalysis, and Sante De Sanctis (1862–1935), who was concerned with psychopathology, introduced both experimental research and practical application. In conservative and Catholic Spain, a number of intellectuals, troubled by the country's lack of openness towards European culture and philosophy in general, adopted the cause of psychology. As the new century opened, scientific psychology had little institutional base and 'men of letters' took the lead in changing the situation. The Catholic response to the natural sciences and to new psychology, however, was not always negative, especially north of the Alps. After the Church's retrenchment against modernity in all its forms in the 1860s, a new pope, Leo XIII, in the encyclical *Aeterni Patris* in 1880, asked for science to be loved for its own sake as a form of truth and called for a revival of Thomist philosophy (the scholarly legacy of Aquinas) to provide intellectual leadership. Within this framework, the Belgian scholar and the figurehead of Belgian dignity under the German occupation in the First World War, Désiré F.-F.-J. Mercier (Cardinal Mercier, 1851–1926), established a Thomist institute at the university of Louvain, or Leuven, in 1889. Here he supported teaching in modern science, which included teaching in experimental psychology. Armand Thiéry and later Albert Michotte (1881–1965) became directors of the Louvain laboratory – the first in Belgium – and they supported both precise experimental work and the connection of psychology to pedagogy. Mercier himself wrote general texts that developed a Catholic philosophical anthropology, defining psychology as 'the philosophical study of life in man' and attacking the mechanist legacy of science attributed to Descartes. Without a philosophy of the soul, Mercier argued, psychology 'is not a separate science, but only a page of mechanics or of physiology'. Michotte was an experimental psychologist who, after training in German experimental work, especially in Würzburg, directed an influential school in Louvain until 1946. He explored the concrete nature of actual experience in studies published as *La Perception de la causalité* (1937; *The Perception of Causation*), where he concluded that we have a direct and irreducible experience of the causal influence of one thing on another. (This means, in Michotte's view, that the relation between sensations is not contingent on association but results from the way the mind intrinsically perceives the world.)

The history of science in Russia is a record of ambivalent identification with the West and pride in a separate path. Government straddled two horses: the one of tsarist absolutism and rule by divine right, the other of modernization that aimed to establish an industrial economy and administer a state capable of competing on the world stage. An administrative, professional and commercial class, with education but no political power, looked to the example of Western European politics and German science. There were colonies of Russian students living abroad; radical, even revolutionary, young Russian women with short hair were a scandal on the streets of Berne or Zurich. With limited but rapid industrialization in the 1890s, Moscow and St Petersburg became huge urban conurbations breeding political agitation, poverty and crime, and educated liberals turned to the social and medical sciences for the tools with which to tackle new conditions. Their efforts included scientific psychology and its application. Early psychiatrists took a particular interest in psychological questions, and it was physicians who introduced German-style psychological instruments. As early as 1885, V. M. Bekhterev (1857–1927) agreed to return from Germany and take a chair at the university of Kazan' on the condition that the university establish a psychological laboratory in the psychiatric clinic. The psychiatrist V. F. Chizh (1855–1922), who had attended both Charcot's lectures in Paris and Wundt's in Leipzig, became an enthusiast for reforming the study of the mentally ill with the help of psychological experiments. When, in 1891, he took over the position of the psychiatrist Emil Kraepelin in Dorpat (now Tartu, Estonia, then the centre of a Russian province), he was able to put his ideas into practice.

The first formal university institute came in 1912, when G. I. Chelpanov (1862–1936), who had also studied with Wundt, opened an institute for psychology, independent of philosophy, in Moscow university. This institute attracted large numbers of students. Chelpanov was an adept organizer and held together an eclectic cluster of interests and tendencies. He hoped that psychology would contribute to a science of education, but when his institute did take an early interest in mental testing, this was as a theoretical activity rather than as a practical intervention in the poorly developed state school system. Russian liberals, it is interesting to note, differed from their colleagues in the West and took it for granted that low performance in intelligence tests reflected the poor social conditions in which the majority of people lived, rather than heredity. There was no institute comparable to Chelpanov's at the university in St Petersburg, and instead a unique organization grew out of medical interests. Bekhterev, who had moved to St Petersburg, was a dynamic entrepreneur, and he managed to put together private and government funds for an independent psychoneurological institute in 1908. It became a large, eclectic enterprise, with different people working on everything from experimental psychology to sociology, and from social psychology to criminal anthropology.

As in Russia, so – in this respect – in Britain: psychology developed piecemeal as a field, with distinctive local characteristics. Spencer, Darwin and Romanes wrote as individuals, not as academics or scientists with a career in institutions. Bain, however, appointed professor of logic in Aberdeen in 1860, became a heavily committed teacher of logic, grammar and mental science. Bain also sponsored the journal *Mind* (founded 1876), which in its first decades had as one of its central ambitions the achievement of a decision about the proper foundations of psychological knowledge. Thereafter, however, it became a journal for philosophy, a discipline, unlike psychology, with an institutional base. In 1892, James Ward (1843–1925) extracted minute funds for psycho-physical apparatus to use in the existing physiological laboratory at the university in Cambridge. In Cambridge and London there was an attempt to bring England into line with current German practice and to study perception as a problem in natural science. In 1912, C. S. Myers (1873–1946) directed a new psychological laboratory in Cambridge, funded largely by himself, and during the 1920s and 1930s this became a centre for training experimental psychologists under Myers's chosen successor, Frederic C. Bartlett (1886–1969). The field slowly expanded in the interwar years.

Around 1900, many British physicians continued to think of themselves as experts in psychology understood as psychological medicine. In marked contrast to the doctors, a number of philosophers, including Ward and his student and colleague G. F. Stout (1860–1944) at Cambridge, and F. H. Bradley (1846–1924) at Oxford, practised psychology as the analytic study of mind. They held different views about the proper terms and subject-matter of analysis, but they agreed that psychology has its own subject-matter and cannot be studied as a function of nervous processes (though they supported study of nervous functions). Bradley berated the empiricists (like Bain) for confusing psychological and philosophical questions, and he thought it evident in reason that the latter must have a foundation in logic not psychology. Nonetheless, he contributed to what he called 'psychology', the analytic description of mental states like cognition and volition. It was Ward who wrote the first entry for 'Psychology' to appear in the *Encyclopædia Britannica*, in 1886. It was an opaque piece, and it did not mention Darwin (except in a passing comment on expression), yet it acquired a reputation for decisively refuting empiricist psychology and instead taking the self as the start and end point of analysis. There was no agreement about analysis, however: Ward and Bradley, for instance, were markedly antagonistic to each other. Stout, who became the second editor of *Mind*, sustained an analytic approach through the first decades of the twentieth century, though his standard *Manual of Psychology* (1899) claimed to set psychology in a developmental perspective. In Britain, in 1901, when a small group of people formed the British Psychological Society, psychology was a varied and divided field.

This variation is even more striking if we consider what was going on in London. In the 1870s and 1880s, James Sully (1842–1923), who then lived off his writing, produced a series of readable books on psychological topics, mixing everyday description, the empiricist conceptions of Bain and Spencer and the evolutionary outlook in general. He promoted psychology as the basis for a science of education, wrote texts used in teacher training, including a study of child psychology, and established a laboratory at University College, London, in 1897. But it was Francis Galton (1822–1911) who was to be the larger influence. This extraordinary man – told as a young child that he was a genius and who, as an older man, was addicted to counting everything from the number of blooms on chestnut trees to 'Ums' in committee meetings – turned to psychology in the light of Darwin's work and understood psychology as above all else the study of individual differences and their place in biological evolution and social progress. This interest, taken into University College by Karl Pearson and Charles Spearman, shaped much of psychology in twentieth-century Britain into the study of intelligence.

This is all very condensed, but it is enough to show psychology's protean character. What psychology *was* depended on local circumstance – on national settings, on institutional arrangements and opportunities, on individual enthusiasm. Educated people interested in psychology, though, all looked to Germany, or at least looked over their shoulders at Germany, where there appeared to be a blueprint for psychology as a rigorous science. Even in Germany, however, there were different claims about the form of the science, as well as questions which arose from social, medical and educational rather than academic debate.

## GERMAN PSYCHOLOGY

In Germany and in the Austro-Hungarian Empire, in 1900, there was specialist activity in psychology but few university positions reserved for psychologists. This apparently paradoxical situation grew out of the way psychological interests developed in the philosophical faculty. Crucial in the mid-nineteenth century was a perception, widespread among the young and intellectually ambitious, that pure philosophy, exemplified by Hegel, had failed to advance in the way the sciences – physics, chemistry, physiology, and also philology and history – had advanced. It appeared necessary either to jump on board the sciences or to adapt scientific methods to renew philosophy. Psychology was the beneficiary, and most directly so when scholars used the methods of experimental physiology for the study of sensation and hence mental processes in general. All the same, philosophy remained of high status and firmly embedded in the institutional fabric of the universities. The result was a tension between those who thought philosophy must have priority as the fundamental discipline and those for whom new experimental methods had proved their superiority as the way to make progress. There was to be no formal separation of psychology as a separate discipline in the nineteenth century, though psychologists did establish specialist societies, journals and training at an advanced level. As we have seen, students from around the world came for this training. What they took back home, however, was what they wanted to take and not necessarily the German philosophical project of which psychology was so often a part. This was notably true for North Americans.

Wilhelm Wundt (1832–1920) established the largest and most visited centre for psychological research, part of his teaching in philosophy at the university of Leipzig. He started out as a medical student but then, especially after studying with the physiologist Du Bois-Reymond in Berlin, saw a medical career as a poor substitute for research. He spent some fifteen years in Heidelberg, publishing textbooks and writing on psychological, physiological and philosophical topics, while hoping to be called to a chair. He was also active in Baden state politics as a liberal. For some of these years he was an assistant to Helmholtz in the latter's physiological laboratory, though there appears to have been little constructive contact, partly due to Wundt's prickly need to establish himself as an independent scholar. Helmholtz was the most brilliant instance of an experimental physiologist who used his expertise to address psychological topics with a rigorous method. Though a natural scientist, he was also aware of the purchase of his work on philosophical topics. It was Helmholtz and his contemporaries in the 1850s and 1860s who laid the basis for what became the experimental form of psychology. Their primary area of interest was sensation, the topic where experimental methods associated with physiology most clearly found application in relation to mind. Helmholtz, Wundt and others established practices which, while firmly acknowledging sensation to be a nervous function, nevertheless maintained *psychological* research with sources of knowledge independent of physiology. This shaped psychology as a separate field. How far their methods were in fact applicable to higher mental activity was to remain a large question.

Among Helmholtz's early work were studies (1850–52) on the speed of conduction of the nervous impulse. Awareness of the measurable time taken by the physical correlates of mental processes then suggested a way rigorously to study the variables affecting mental life, using time taken as a sign of activity. Astronomers had been aware from the end of the eighteenth century that different observers record the moment a star crosses a hairline in a telescope slightly differently. The difference was regular for a particular observer. The phenomenon appeared initially to be due to observational error, but psychologists re-analysed it as a personal equation, an individual variable in the sensation of a visual stimulus. An influential paper in 1862 by the Dutch researcher on perception, F. C. Donders (1818–1889), then pointed towards the general value of reaction-time experiments, and a swathe of reaction-time studies followed. Given what appears (to me at least) their staggering tedium, we may take the effort invested in them to be an index of the enthusiasm aroused by hopes for a breakthrough to scientific study of mind. It was another matter, however, to establish clear and authoritative ideas about *what* was being measured.

Instruments had an important role in these studies. There was an intimate relationship between what instruments made it possible to observe and measure and what sort of research scientists undertook. Helmholtz, who worked in physical as well as psycho-physiological optics, developed the ophthalmoscope to look at the retina at the back of the eye. Mathias Hipp, a watchmaker, made a chronoscope in the 1840s (an instrument able to measure time intervals to a thousandth of a second), and without this reaction-time research would have been inconceivable. The ability to use such instruments became a hallmark of the trained scientific psychologist, and this skill separated the scientist from the armchair psychologist – the 'amateur'. Indeed, instrumentation was such a feature that later observers referred to 'brass instrument psychology', and for sceptics this was more of a compliment to the proficiency of German precision engineering than to the psychological results. The use of instruments made psychology a scientific occupation even if it did not create scientific knowledge. In the area of sensation, however, precision tools were appropriate for the questions asked.

Helmholtz published a massive three-volume *Handbuch der physiologischen Optik* (1856–66; *Handbook of Physiological Optics*) and *Lehre von den Tonempfindungen als physiologischer Grundlage für die Theorie der Musik* (1863; *Course on the Sensations of Sound as Physiological Basis for the Theory of Music*). These became standard reference points in their field and demonstrated how experimental methods and physiological concepts could be put to work on psychological topics of visual and auditory perception. Helmholtz's debate about vision with Ewald Hering (1834–1918) linked experimentation on sensory perception and philosophical theories of the sources of knowledge, as experimentalists believed that their work addressed in scientific terms the questions bequeathed by Kant about how the mind can know the world. Helmholtz and Hering disagreed about innate organization within the brain and its role in perception. Helmholtz argued that perceptual experience involves the mind forming elementary sensations into a unity. To some extent influenced by Kant, he maintained that perception involves central psychological processes in addition to the sensory data, and he called these mental processes 'unconscious inference'. He believed, for instance, that visual perception of depth is an acquired perception achieved by mental unconscious inference from sensations provided by binocular vision (in which the two eyeballs move slightly as they focus on different objects in the visual field). His position was therefore described as empiricist. Hering opposed all this and argued that an in-built structural organization explains perception. His position was therefore described as nativist. In reality, a confusing mix of psychological and physiological language bedevilled perception studies, as the term 'psycho-physical' research indicated.

The Helmholtz–Hering debate, which continued in altered form into the 1920s and beyond, exemplifies how philosophers and physiologists constructed psychology as an academic natural science subject. In spite of philosophical differences of opinion, the laboratory became the site where differences were to be resolved. Younger scientists, who were caught up in an intense competition to obtain permanent positions, used the laboratory and their experimental training to make their mark. The hope that studies of perception would also transform matters of dispute in philosophy made such research doubly attractive.

After the publication of Gustav Theodor Fechner's *Elemente der Psychophysik* (1860; *Elements of Psycho-Physics*), there was a field specifically known as psycho-physics. Even among German professors, Fechner (1801–1887) was an unusual man. His career began with a chair of physics in Leipzig, where he wrote satirically about natural philosophy under the pseudonym Dr Mises. He proved, for example, that as angels are the most perfect beings they must be spherical. He then ruined his sight in optical experiments, went through a spiritual crisis and resigned his chair. On his recovery, he developed a world view that pictured the mental and the physical as alternative representations of one reality. In his *Zend-Avesta* (1851; the title he borrowed from the sacred text of Zoroastrianism), Fechner described the universe as a being with consciousness. He was therefore centrally interested in the unity of the material and the mental, and in experiments on the relation between the changing physical stimulus and the changing mental perception of the change he claimed to find a constant mathematical correlation. The model experiments were those of E. H. Weber (1795–1878) in the 1830s on the just noticeable difference in discriminating weights. Fechner thought that by using a subject's report of just noticeable difference one could map subjective sensation against the objectively measured sensory stimulus. The claim to have found a mathematical constant relating the mental to the physical generated excitement and the field of psycho-physics. It appeared to initiate a general programme making mental events open to precise description through their correlation with measurable physical changes. If few researchers accepted Fechner's results in precisely the terms in which he stated them, he nevertheless launched a topic which shaped scientific psychology for a generation. With institutional support, the relevant techniques became the core of a self-perpetuating specialist discipline; in the process, questions to which answers could not be formulated in quantitative terms tended to get deleted from the agenda. Fechner himself, though, turned to pioneering work in experimental aesthetics.

Wundt's early experimental research was on the physiology of muscles and nerves, though his ambition went far beyond this and encompassed an evolutionary perspective. Academic success came with the publication of his *Grundzüge der physiologischen Psychologie* (1873–4; *Principles of Physiological Psychology*) and his call in 1875 to one of three prestigious chairs of philosophy in the university of Leipzig. The philosophers in post wanted their subject to engage with the pacesetting world of the natural sciences. Wundt responded over the next 40 years with what, in intent if not in result, was a unified philosophy made possible by natural science methods deployed in psychology. He did not advocate psychology as a subject distinct from philosophy, though he taught specialized psychological methods; and he increasingly differentiated psychology from physiology, though physiology had provided him with his experimental orientation. He had a definite programme for psychology, with a claim to leadership presented in successive editions of *Grundzüge* and with ample resources in Leipzig, and he therefore became a focus for international debate about what sort of science psychology should be.

Wundt did not claim, as some of the physiological psychologists discussed earlier did, that knowledge of the nervous system, which he outlined in his text, made psychology into a science. Psychology, he stated, has its own methods and content. There is, first, experiment, and he taught the techniques of reaction-time studies and psycho-physics. The techniques, he argued, do not involve introspection but rather the correlation of reported change in the size, intensity and duration of sensations with physical variables. For example, subjects released a key in an electric circuit at the moment they perceived a change in the brightness of a measured light source. The experiment could be repeated as often as needed in order to get regular results. Wundt set aside time in the curriculum for a practicum, or period of laboratory work, and he trained advanced students to work up theses using his methods. Students employed each other or even Wundt himself as experimental subjects since they supposed that only experienced subjects properly carry out the tasks. Experimenter and subject also changed places: the purpose was to study general mental activities not personal qualities. Wundt secured the institutional support necessary for the expansion of his programme (for funding instruments, space and assistants) and he founded a journal, *Philosophische Studien* (1883; *Philosophical Studies*), in which to publish results. His efforts culminated in an institute in 1883, which expanded in 1897 into a purpose-built centre. The message got around that training in scientific psychology could be acquired from a period in Wundt's classes and practicums.

It nevertheless remained Wundt's intention to develop psychology as a science contributing to the solution of philosophical questions. He held mental action to be a reality in its own right, and it followed that psychology must be an independent science and not a branch of physiology. Psychology, he thought, uncovers the laws of psychic causality, and the preliminary step is the accurate description of mental content. Wundt classified action into three forms: impulsive acts (caused by *Trieb* – living impulse or instinct, rather than 'drive'); voluntary acts, in which one motive dominates over others; and selective acts, in which one motive dominates after a conscious choice. He linked this hierarchy of actions to the evolutionary hierarchy from animal to man. In practice, it was difficult to imagine ways to examine volition experimentally, and Wundt himself excluded the will from the scope of experiment and discussed it analytically, rather in the manner of Ward or Stout in Britain, as part of mental activity in which the mind judges.

Wundt lectured, published and supervised students on logic, ethics and the history of philosophy, as well as on psychology. At base, he addressed psychological questions in the light of his belief in the mind's rational activity as the precondition of knowledge. He did not think this rational activity, any more than volition, could be the subject of experiment. Indeed, large parts of his programme for psychology involved non-experimental methods. To examine complex mental activity, including thought and language, the feelings and volition, he turned to a kind of cultural psychology widespread in German-speaking Europe. With roots in Herder's late eighteenth-century vision of the human spirit expressing itself in history, this way of thought looked to culture and language as the expression of mind and hence the source of knowledge in psychology. Wundt appropriated the word '*Völkerpsychologie*' (psychology of the life of peoples, or 'ethnopsychology') to denote psychology which studies mind through interacting minds, mediated by language, thus creating the social world of myth and custom.

This side of Wundt's interests dated from his early work on developmental principles, which included the idea of *Trieb* as a basic affective process leading to expressive movements or gestures. Gestures are replicated by others, hence communication and shared mental states become possible, and when refined, this communication becomes language. Shared mental states constitute a *Volksseele* (soul of a people), a trans-individual psychological structure which, over time, creates culture. Beginning in 1900, Wundt published a series of volumes taking this approach to the psychology of the higher mental processes. For reasons which will become clear, this work had a much smaller audience among the younger generation in search of a scientific psychology than the experimental research. Nevertheless, the notion of gestural communication – Wundt's view of how the individual mind creates a shared culture – was taken up in social psychology (for example, in the work of G. H. Mead).

In historical circumstances where religious authority was in question and in which many people perceived the churches to be hand in glove with political reaction, serious young students turned to psychology as a subject at the forefront of objective studies of the human condition. They were naturally attracted to Leipzig. Nevertheless, there were other centres of psychology in German-speaking countries and other views of the science; indeed, there was intense competition.

G. E. Müller (1850–1934) at the university of Göttingen was Wundt's main rival in training students in experimental technique. He was the successor to the philosophy chair of Rudolf Hermann Lotze (1817–1881), who had written on natural science and the philosophy of mind and whose synthesis for a general audience, *Mikrokosmus* (1856–64), related human values to cosmic reality. Lotze was a voice for constructive dialogue, as opposed to polarization, between younger natural scientists and older philosophers who feared the mechanist views which natural science appeared to be constructing. Müller, by contrast, took for granted the experimental method in psychology and restricted his work almost entirely to narrow questions, such as the effect of different sensory conditions on memory. He did not pretend to have a comprehensive programme to unify knowledge like Wundt but gave undivided attention to experimental activity. Müller, not Wundt, was the guiding spirit in the foundation of the German society for experimental psychology in 1904.

Whatever Wundt's position, it was only a matter of time before others made the attempt to adapt methods used in reaction-time and psychophysics research to study the higher mental activities of attention, memory and thought. Hermann Ebbinghaus (1850–1909) took this step when he devised a way to study memory experimentally. Working on his own initiative, outside an academic setting but inspired by Fechner, he invented original ways to study memory (or what he took memory to be). He explored his own ability, varying the conditions, to memorize lists of nonsense syllables and cantos from Byron's 'Don Juan', measuring the amount of material remembered over time and plotting the results graphically as a curve for forgetting. When he published *Über das Gedächtnis. Untersuchungen zur experimentellen Psychologie* (1885; *On Memory: Investigations on Experimental Psychology*), it made his name because readers immediately recognized, even in the title, that it opened a road to the higher, cognitive activity of mind. Ebbinghaus reinforced the message with a provocative epigram in Latin, a manifesto assigning the future to psychology and the past to philosophy: 'From the most ancient subject we shall produce the newest science.' He went on to teach in Breslau and Halle, to co-found the *Zeitschrift für Psychologie und Physiologie der Sinnesorgane* (*Journal for Psychology and Physiology of the Sense Organs*) in 1890 and to publish a textbook.

Though Wundt ruled out an experimental approach to complex mental processes, his students and assistants began to go their own ways, and by 1900 there was competition between alternative schools of thought. The most important new group was headed by Oswald Külpe (1862–1915), who studied in Berlin and with Müller in Göttingen before he completed a thesis with Wundt. He published on reaction-time experiments but then concentrated on philosophy and aesthetics after he became a full professor at the university of Würzburg in 1894. In Würzburg he supported a more expansive view of experimentation which, for example, involved subjects in making complex introspective reports, and the large amount of experimental work coming from his laboratory changed views on perception and thought considerably. There were sharp exchanges between Wundt and those in Würzburg on applying experiments to thought itself and on the nature of introspection. Participants began to acknowledge how extraordinarily difficult it was, and some were to think impossible, to devise experiments which would persuade other researchers to accept the results. Different research groups simply could not agree about what they studied in the mind.

One body of opinion maintained that the issues were at least as much conceptual as experimental. This once again posed the question of psychology's relation to philosophy, which was at the same time a sensitive matter of employment and status. As in Britain and France, there was emotive debate about the proper spheres of philosophy and psychology. German psychologists certainly established the elements of a discipline (specialist training, journals, institutes) between 1880 and the First World War. At the end of this period, there were perhaps ten experimental psychologists who were full professors, but all of them occupied positions in philosophy. Their main student audience consisted of future *Gymnasium* (classical secondary school) teachers, but their main academic peers were philosophers, who were increasingly concerned about the number of experimental psychologists who had replaced philosophers in posts and teaching. Besides the question of posts, argument about the relations between the disciplines signalled different conceptions of the relations between logical reason and empirical science in grounding knowledge.

Ernst Mach's philosophy of science suggested a way forward and influenced many natural scientists. Mach (1838–1916), who taught at the university of Prague and then in Vienna, rejected physical concepts if they could not be re-expressed as relations between sensory elements. (The commitment to absolute space-time in Newtonian mechanics was a particular target.) Similarly, he criticized psychological concepts (like Helmholtz's 'unconscious inference') that were not derivable from or translatable into the relations of sensory elements, while also carrying out experimental studies on space perception. His philosophy of knowledge, positivism, cut out ordinary language, which represents a self perceiving the physical world, in order to describe regularities in phenomenal consciousness. Mach thought this step relegated the mind–body problem and concern about the relation of psychology to physiology to the status of non-issues: 'There is no rift between the psychical and the physical, no inside and outside, no "sensation" to which an external "thing" different from sensation, corresponds. There is but one kind of elements.' Psychology, it followed, is the basic discipline in the theory of knowledge, since it provides descriptions of what the elements are, the 'positive' facts, which constitute the world. Philosophy after Locke, which had analysed knowledge into ideas with a sensory origin, was of course precedent for this approach, but it was German experimental technique which made the approach a truly empirical one. Just like the earlier tradition of thought, however, Mach faced critics who judged the whole endeavour ill-founded, since there did not appear to be empirical support for the basic assumption that sensations exist as atomistic elements.

Mach's argument attracted the young Külpe. Nevertheless, later work in Külpe's laboratory showed Mach's philosophy to be unviable. First, experiments indicated the way thought involves mental states, such as attitude, which are not sensory in nature. This precipitated a debate about what in English was called 'imageless thought'. Second, Külpe and some of his associates concluded it was better to describe the content of mind in terms of mental *acts* than in terms of mental *elements*, sensations or presentations (*Vorstellungen*). Debate about these topics was to persist, indicating psychology's closeness to philosophy rather than its emancipation as a separate academic subject, however much psychologists developed special expertise.

The language of mental acts drew on yet another literature on psychology, associated with the German philosopher Franz Brentano (1838–1917), who lectured in Vienna. His significantly titled book, *Psychologie vom empirischen Standpunkt* (1874; *Psychology from an Empirical Standpoint*), stressed the empirical source of knowledge but did not adopt the new experimental methods. Brentano's aim was to make psychology empirical through the accurate description of conscious reality and not by the establishment of a new natural science. He argued that the mental, the subject-matter of psychology, as opposed to the physical, is by definition an act not a state. For the purpose of analysis, he supposed, mental acts have two objects: the immediately observed primary object (e.g. the teapot) and the perceived but not immediately observed secondary object (e.g. the seeing of the teapot). The distinctive business of psychology, Brentano therefore argued, is 'inner perception' of the secondary object. Though it requires special attention, inner perception is the means by which we gain knowledge of what makes the mental life of human beings a *human* life. For Brentano, mental activity was irreducible, not analysable into something else, and his aim was to establish empirical psychology in order to describe this activity rather than the physical or other primary objects to which mental activity is directed. In technical terms which derived from medieval Thomist philosophy, he defined the mental by intentionality: the mental is an act about or towards an object. This was suggestive to psychologists who were under pressure to say why their field should be independent from physiology in one direction and philosophy in another. The language of acts and intentions described psychological life in terms of its own nature, and it did not describe it in terms of physical states, as studies in physiological psychology appeared to do, or leave it without content, as purely philosophical analysis threatened to do. The Würzburg researchers found use for this kind of language in a mass of experimental studies.

Brentano's philosophy had a direct influence on experimental psychology through Carl Stumpf (1848–1936), who was for a time Brentano's student. Stumpf, not the pioneer Ebbinghaus or the experimentalist Müller, was the university of Berlin philosophy faculty's choice in 1893, when it created a position for the new psychology. It seems likely that the faculty approved of Stumpf's conviction that neither Wundt's *Völkerpsychologie* nor the experimental analysis of thought could displace reason's description of itself as conscious activity. Stumpf certainly held sophistication in philosophy to be a condition of psychology's seriousness as science: 'The psychologist must at the same time be a theorist of knowledge . . . as anyone must for whom science is more than artisanry.' In Berlin Stumpf created a major centre for experimental psychology, but he did not make psychology an independent natural science. Rather, his psychological seminarium and the experimental work it supported, refined a philosophical science committed to clarifying, in its own terms, knowledge as it appears in consciousness. This, as discussed later, supported the development of phenomenology.

Stumpf's vocation was philosophy, his rational conviction psychological experiment, and his passion music. He played the cello while he studied and taught in a number of universities, and music and psychology came together in his major study on *Tonpsychologie* (1883–90; *The Psychology of Sound*). Once in Berlin, he successfully organized facilities for psychology, including laboratory space, but he handed experimental work over to others. His students included Wolfgang Köhler and Kurt Koffka. Together with Max Wertheimer, they formed what became known as 'gestalt psychology' – a major influence on theories of perception in the twentieth century – and Köhler succeeded Stumpf on the latter's retirement. Their work perpetuated the close association of experiment and philosophy; in this respect like Wundt, they hoped psychological research would make it possible to found philosophy as science.

An influential member of the faculty which recommended Stumpf was the philosopher Wilhelm Dilthey (1833–1911). Histories of psychology tend to ignore him, albeit he had an agenda for advancing psychology. From the viewpoint of experimental psychologists, it did not count because it raised large questions about the whole project to which they were committed, and this made Dilthey appear not relevant. From a viewpoint which does not assume it is a cut-and-dried matter to know what makes psychology a science, however, his writing is of great interest. It is possible to read Dilthey for his commentary on the nature of knowledge in psychology in connection with history and culture and thus understand the disunity and insecurity of psychology as a subject area at this formative period. Moreover, he used terms and concepts which have continued to inform discussion of relations between the natural sciences and other branches of learning, discussion in which psychology has had a pivotal position. But there is no gainsaying the fact that he was a wordy and at times difficult writer whose arguments revealingly shifted. Simple summary is difficult.

Dilthey was duly impressed with the natural sciences, and in his early research on literature and biography he sought to make such work equally scientific, not just dependent on intuition and knowledge of 'the human spirit'. His intention was to achieve objective knowledge in *die Geisteswissenschaften* (literally, the sciences of spirit, roughly corresponding to the modern field of the humanities) with a method for understanding the historical context of both subject-matter and the interpreting observer. With knowledge of what a past text meant (to author and reader) in its own time, and knowledge of the interpretative framework and values of the scholar (who for this purpose must become self-reflexive), it should be possible, so to say, to step outside all contexts and view objectively. He looked to psychology to become a science competent to contribute to this method, publishing *Ideen über eine beschreibende und zergliedernde Psychologie* (1894; *Ideas Concerning a Descriptive and Analytic Psychology*) in support. We can, Dilthey supposed, understand and enter into the psychological relation of motive and action in another person or historical figure because minds share a common basic structure. Understanding this structure is necessary for interpretation, the science of meaning. Through psychology, we make intelligible other people's as well as our own reasons and actions, and psychology as a science will be systematic knowledge of the purposeful action of people making the world around them. The problem, as Dilthey himself became increasingly aware, was that it did not appear possible to achieve an absolutely objective stance, outside history as it were, as the basis for an unquestionable understanding of human life. Even with reflexive self-awareness, the interpreter (historian, literary scholar, psychologist) brings values into the interpretative process. For Dilthey, as for other German professors, 'mandarins' of the nation's cultural inheritance, values and their expression in life are or should be the ultimate subject-matter of the human sciences. But from whence an objective science of values? This question, at the end of the nineteenth century, gave rise to a considerable literature voicing a sense of crisis in the intellectual path of the sciences, psychology included. Faced by many criticisms, Dilthey questioned whether psychology could achieve a universal standpoint. Instead, perhaps, he implied, knowledge of being human involves a continuous reappraisal of our situation and ourselves. This pointed towards a future for psychology among the disciplines of the humanities, like literary theory, rather than among the natural sciences.

One simplifying response to these difficulties was to take the distinction in German between *Naturwissenschaft* (natural science) and *Geisteswissenschaft* and erect it into a basic distinction between different kinds of knowledge appropriate for different subject-matters: the former for nature, which just 'is', and the latter for culture, which displays 'values'. Some modern writers do this in order to differentiate between the natural and the human sciences, and Dilthey sometimes came close to making the same move. Thus, he argued, to understand nature is to describe law-like causal relations; by contrast, to understand human beings is to recreate the mental world, the meanings and reasons, which inform human actions. He claimed: 'We explain nature; we understand psychic life.' This division between causal explanation and understanding was then, and again a century later, central to debate about relations between the sciences and the humanities. The varieties of psychology in 1900 occupied, as they still occupy, positions on both sides of the division. This is one reason why I am keen to recognize historical variety.

Another simplifying response was to hold up the natural sciences as the uniquely objective source of knowledge. There were a few people who did want to argue this position (as some neuroscientists do now), and Dilthey, like most other German professors, was utterly opposed. This was because the human sciences, including psychology, have the expression of values in culture as their subject-matter. Following Dilthey's 1894 study, and following Dilthey's opposition to Ebbinghaus's application to the Berlin chair, Ebbinghaus attacked Dilthey and affirmed that psychology is objective only in as far as it is an experimental natural science. This and the bad-tempered exchange which followed showed just what a divided field psychology was. Psychologists who have published histories tend to assume Ebbinghaus was obviously right and have therefore written Dilthey out of the story as no psychologist at all. In doing so, however, they express the *value* of psychology being thought of as natural science, and it is exactly the expression of such values, Dilthey believed, we should understand if we want a science of people.

It is easy to get the impression that every German academic was a closet philosopher. This was not so: Müller, a strict experimentalist, was an outstanding exception among psychologists. More importantly, there were numerous psychologists, teachers and social scientists who looked to psychology to make a difference to education and social administration. A prominent example is the physiologist, William T. Preyer (1841–1897), who broadened his work from studies of reflex control and psycho-physics to discuss perception, linguistics and child development. In *Die Seele des Kindes* (1882; *The Soul of the Child*), he combined physiological, evolutionary and psychological information with the clear purpose of contributing to practical life. He also developed recommendations for the reform of the Prussian state school system. Many teachers, in Germany as elsewhere, were interested in the possibility of psychology providing a scientific basis for education, and their expectations put quite different pressures on psychology from those originating in academic philosophy. If we contrast Preyer and Wundt, this can clearly be seen. Preyer was interested in the individual person linked by evolutionary history to nature and by education to society; Wundt was interested in the universal characteristics of the conscious mind and its expression in linguistic and cultural life. Preyer's subjects were children, and Wundt's his students and colleagues.

Many more psychologists became interested in psychology as a practical occupation in the early years of the twentieth century. One of Wundt's most successful students, Ernst Meumann (1862–1915), shed his teacher's reservations and directed work in experimental educational psychology. He edited journals on 'general psychology' and 'experimental pedagogics', terms indicating the degree to which hopes for practical psychology took psychologists well beyond Wundt's programme. William Stern (1871–1938), who became director of a new psychological institute in Hamburg in 1916, was interested in a great range of experimental and practical topics from the beginning of the century, when he taught at the German university of Breslau (the Polish city of Wrocław after 1945). He hoped that psychology would contribute to the problems of industry, and he contributed the term 'intelligence quotient' to the literature on children's school performance. Nazi race law removed Stern from his post in 1933, even though for a short time Stern had had the illusion that the new regime, which was committed to modernization, would value his ideas for human management. Yet, strikingly, in spite of these interests, Stern made no attempt in his teaching to separate psychology from philosophy.

Hugo Münsterberg (1863–1916), another of Wundt's students (though Wundt did not approve of the work he did on volition), and who was persuaded to move to Harvard by James and by the anti-Semitism blocking appointments in German universities, became a leading advocate of the application of psychology. Like Stern, Münsterberg believed in the relevance of psychological research, especially on perception, and he was interested in industry and the legal issue of witness testimony. During the first decade of the twentieth century, he and other psychologists, supported by reform-minded lawyers, placed high hopes on the development of legal psychology. These hopes, however, did not survive the rigours of the adversarial process in U.S. law, which was apt to expose weaknesses in psychological claims. His advocacy of psycho-technics – the use of instrumentation to describe and classify human performance in general and individual differences in particular – had more success. It appeared directly relevant in the practical world where people operated machines and drove trams.

The history of German psychology exhibits two things. First, there was great diversity, and even within the restricted compass of experimental psychology, different research groups did not unite around a common programme. Second, psychology did not take off as an independent discipline. Nearly all the academics who called themselves 'psychologists' held appointments in philosophy, and most senior psychologists thought this right and proper. There were those, nevertheless, like Preyer, Meumann and Stern, who linked psychology to child development and education. The situation changed slowly during and after the First World War, when both the military and industries, like the railways, showed an interest in the rationalization of their activities and began to invest in psycho-technics. Practical psychology was taught quite widely in technical universities during the 1920s, and there were six psychology chairs in German universities by 1930. Then a large-scale change in the fortunes of psychology as an occupation came during the period of the Third Reich. In the late 1930s, psychologists claimed expertise in personnel selection, and the military interest in this made available the resources to establish a distinct occupational identity.

## PSYCHOLOGY IN THE USA

Psychology had a more defined shape in the United States than anywhere in Europe at the end of the nineteenth century. It had a professional organization, journals, a substantial place in the universities and a clientele willing to consider its claims for funding and practical utility. The subject existed separately from philosophy, and if the psychologists had concerns about intellectual boundaries and status, they were about psychology's autonomy from physiology: it was not self-evident why psychology, if it were a natural science, should not be a branch of physiology.

The uptake of experimental methods from Germany, clinical material from France and evolutionary thought from Britain was rapid and thorough. This did not involve a complete break with the earlier teaching of psychology as part of moral philosophy, but, rather, the earlier moral content reappearing in claims for the *social contribution* new psychology would make. McCosh at Princeton, though he lectured on motivation and emotion as part of Christian moral teaching, was notably open to both evolutionary theory and German psychology. Differences emerged when enthusiasm for psychology as natural science appeared to Christians to tend towards materialism, and, of course, there was conflict when local communities strongly held to the Bible and the word of God as the only foundation for the conduct of life. Students who wanted to take their education further, encouraged by their colleges, travelled to Europe, especially to the German universities. From the 1870s, these students included some who were dissatisfied with moral science as a science of the mind and who looked to experimental science for a more modern and objective way to study people. They naturally went to Leipzig, though not only to Leipzig, as the most famous centre for psychology. When they returned home, they returned to a vibrant educational scene with funding possibilities and institutions open to innovation.

From the 1860s, the country's western expansion, civil war, industrialization and the flood of immigrants from Europe, and to a lesser extent Asia, caused old moral and cultural certainties, taken for granted in the lives of an East Coast elite, to lose authority. New private universities like Cornell, Stanford, the Johns Hopkins and Chicago opened their doors and began to incorporate the German model of the research disciplines. The older private colleges, like Harvard, set about wholesale reform of curricula in response to new conditions. Meanwhile, in 1862, the Morrill Act made land grants available to state legislatures, and this led to the foundation of state colleges with close connections to local businesses and social reformers. These colleges had to demonstrate their usefulness to the communities which funded them. The overall result was a major expansion of the higher education system, in which, in spite of tensions between the interests of academics and local people, specialist scientific disciplines grew up around postgraduate teaching schools. This was the setting, at the end of the century, in which there was the world's first large-scale investment in psychology as a distinct discipline.

One of Wundt's earliest American visitors (though not a student) was G. Stanley Hall (1844–1924), a key figure in founding the psychology profession and academic departments. Like a significant number of other early North American psychologists, he came from a religious family and had an early career trajectory which looked as if it would take him into Christian ministry. Not without reason, observers in the twentieth century were often to perceive psychology as a replacement for religion, perpetuating moral concern with individual achievement and human relations but turning to modern objective methods rather than ancient words for authority. After a long struggle to establish himself and to shake off the odour of materialism attaching to scientific psychology, Hall was instrumental in founding Clark University in Massachusetts (1887), where he became both president and head of a separate department of psychology. He focused on children's development and education, and in manner and content this interest revealed the continuity of moral purpose between old and new psychology. Hall went so far as to claim that the Bible 'is being slowly re-revealed as man's great text-book in psychology'. He was the leading organizer of the child study movement and he wrote on *Adolescence* (1904), which is widely credited with consolidating this period of growing up as an event to be named, studied and moralized over in its own right. He straightforwardly told educators in 1887: 'Pedagogy is a field of applied Psychology.'

James McKeen Cattell (1860–1944) was the first student to study at length with Wundt and to spread experimental psychology in North America. His biography well illustrates the shaping of psychology in the U.S. and the changes which took place in translating psychological ways of thought from the German context into a different society. Cattell (recall Hall) came from a Presbyterian background, reinforced by the courses he took in the liberal arts at Lafayette College in Pennsylvania, where his father was the president. He then studied in Göttingen, Leipzig and the Johns Hopkins in Baltimore. Cattell returned to Leipzig in 1883 to work on his doctorate and for three years he was an assistant to Wundt, valued for his skill with instruments and at recording reaction-times. Not least, he introduced Wundt to the typewriter. He then worked in England, set up psychological instruments at the University of Cambridge and acquired a strong interest in individual differences from Galton. In 1889, Cattell became professor of psychology at Pennsylvania, where he established a laboratory and a programme to make mental measurements; in 1891, he moved to Teachers College at Columbia University in New York to repeat the pattern. He stayed at Columbia until 1917, when he retired after he had upset many people, both for his pacifism and for his attacks on control over the curriculum and research by college presidents and trustees.

Cattell's programme for mental testing in the 1890s showed what sort of European psychology crossed the Atlantic and what did not. He used his instruments to measure individual differences in sensory capacity and motor performance, hoping to reveal correlations. There was a flurry of interest in the techniques as possibly useful to education by providing physical measures of brightness or dullness in pupils. Joseph Jastrow

(1863–1944) headed a team at the Chicago World's Fair in 1893 that carried out a mass of measurements on American and foreign visitors, following the example of Galton who had set up an anthropometric laboratory in London. Cattell's and Jastrow's objectives were factual and utilitarian. Though the programme of measurement was a scientific failure, as it demonstrated no correlations or insights of lasting significance, it did detach psychology from a highbrow philosophical context, so evident in Germany, and relocated it within a setting in which, at least in theory, it had value to everyman. E. W. Scripture (1864–1945), who took over the psychology laboratory at Yale College in 1892 after he had trained with Wundt, was openly antagonistic to philosophy. His popular books, he wrote, are 'expressly for the people . . . as evidence of the attitude of the science in its desire to serve humanity'.

Cattell taught none of Wundt's philosophy, showed no interest in his *Völkerpsychologie* and unambiguously treated psychology as a natural science founded on the collection of facts. In his background, education served Christian social progress, and he assimilated in Germany only what appeared relevant to psychological expertise applicable to social needs. He was more excited about Galton's techniques for the study of individual differences than he ever was about Wundt's philosophy of mind. Once settled in the U.S., Cattell became an entrepreneur on behalf of research on human capacities thought relevant to social performance. He helped establish the *Psychological Review* and increasingly used his organizational and editorial experience on behalf of journals concerned with the public understanding of science and with science education. In the 1920s, he was active in the Psychological Corporation, a commercial enterprise to market psychological expertise to business and the public. Finally, he was for many years a moving spirit in the AAAS, the American Association for the Advancement of Science, the public forum for the national scientific community.

Like many in their generation, Cattell, Jastrow and Scripture believed in expertise as the means to make the country modern, civilized and democratic. They were devoted to science as an occupation which would contribute to social advance rather than to learning for its own sake. Precisely because this attitude was widely shared, psychology quickly acquired an academic base in the U.S. in the 1890s. Academic staff, administrators and funding bodies all had hopes about what science and education could do for the American people. Young scientists bred on German technique but born with American values put forward a new discipline and offered their services to make this discipline work. They promised to show how knowledge of individual psychological capacities would smooth the birth of a modern society. Given the huge social transformation then going on, with the shift from rural to urban life, the creation of modern business practices and mammoth inward migration, this was an offer hard to resist. There was an urgent need to equip immigrants from rural states and from Europe with the tools for modern urban life. There was also belief, which Dewey so strongly upheld, in a democratic politics requiring individual fulfilment through educational opportunity.

Those who talked about democracy did not necessarily believe in it as far as women were concerned, though Dewey did. New Zealand was the first country to introduce votes for women in 1893, while the USA followed suit only in 1920. The women's campaign in Europe and North America for education and training to enter the professions, especially medicine, had moved onto the political stage at the end of the century, but the fact remained that even a highly educated woman needed a determined character to gain professional acceptance. Psychology attracted women, for example Margaret Floy Washburn (1871–1939), the first woman to receive a PhD in the field (from Cornell in 1894), and a proponent of a motor theory of cognition, and Beatrice Edgell (1871–1948) in London, who combined the British interest in philosophical psychology with setting up a laboratory, having trained in Würzburg (where she was reportedly the first woman to receive a higher degree). Mary Whiton Calkins (1863–1930) earned a much-lauded PhD at Harvard in 1895 but the regulations of the college refused the qualification to women. She went on to establish a laboratory at Wellesley College and to focus psychology on the self. Later, women in large numbers trained in psychology, and their presence was to be a characteristic of the field, more, though certainly not exclusively, in the practical than research areas. Psychology, often hand in hand with education, made a major contribution to the twentieth-century shift in occupations central to women's growing social equality.

Psychologists thought claims for social usefulness and scientific status were two sides of the same coin. James had mischievously responded to critics who described his *Principles of Psychology* as insufficiently scientific by saying that it was 'a mass of phenomenal description, gossip and myth'. But most psychologists were unhappy with such a self-description. They felt vulnerable sandwiched between natural scientists who argued that as psychology becomes scientific it becomes a branch of physiology, and lay people who felt psychology was simply common sense.

In these circumstances, it was important for psychologists to be able to claim a unique expertise, the study of human capacities, which appeared relevant to a major slice of social life, namely, education. This explains the emphasis on psychological tests. The largest single constituency for psychological expertise was teachers, public authorities and those with the funds to invest in education. In consequence, psychologists who started off with narrowly focused experimental studies, for example, on perception, tended to move towards less rigorous but more obviously relevant work on child development or aptitude testing. This was in order to demonstrate what their subject had to offer. Once firmly established in the university structure, though, psychology departments trained an audience sympathetic to their own activities and acquired an institutional momentum of their own. This made possible a considerable amount of experimental research on the German pattern. When their subject had become institutionalized, psychologists found it easier to argue for the status of their subject as science.

A commitment to functional explanation, relating mind to adaptation, had a significant place in mediating between conceptions of the field as a science and as a social utility, between academic values and public values. This was especially visible in the University of Chicago, an institution which had great influence on the psychological and social sciences. Endowed in the 1890s with funds from John D. Rockefeller, it brought together an outstanding faculty in order to give the city a reputation for culture and to address the city's fearsome social needs. Chicago had the nation's social problems writ large: large immigrant groups, many without the English language; exponential urban growth and housing shortages; confrontational labour relations; politics dominated by local bosses; poverty and crime. In these circumstances, it was not hard to argue for education and expertise as the means to achieve social integration. Just as the new steel frame buildings and early skyscrapers created a modern urban landscape, so, it was hoped, the psychological and social sciences would create the fabric of modern men and women.

Dewey became a professor in Chicago in 1894 and under his aegis both experimental and educational psychology expanded. He brought Mead with him and then appointed James R. Angell (1869–1949), who had studied at Harvard under James. These academics shared a functionalist orientation: their interest was the whole person in her or his adaptive relation to the social environment. Faith in science as utilitarian expertise, as the highest stage of evolution, permeated their work. At the same time as Dewey organized the philosophy department to make space for a socially oriented psychology, Albion Small created a large and influential department in the social sciences with an interest in the psychological dimensions of human relations. And we may just note that, in Chicago, J. B. Watson wrote a thesis on the psychology of rats and began to think of psychology as a technology for the control of behaviour – a goal which he and his successors saw as the culmination of the evolutionary process.

Psychology therefore developed in the U.S. in a different way and on a different scale from elsewhere. If German psychologists had to justify their existence to philosophers and to conservative state politicians who had the final word on appointments, psychologists across the Atlantic had self-styled practical men to satisfy. All the same, just as practical psychology also developed in Germany, so academic psychology also expanded in the U.S. Edward Bradford Titchener (1867–1927) was a significant figure in this regard. Titchener was an Englishman, more impressed by physiology than philosophy as an undergraduate at Oxford, who wanted to study mind scientifically and therefore went to Leipzig to write a thesis on reaction-time experiments. In 1892, he went to Cornell to head a recently opened laboratory, and there he stayed. Cornell's president, A. D. White, had fought an extended but ultimately successful battle against the religious interests dominating college life. It was this battle, projected onto the past as a whole, which inspired White's book *A History of the Warfare of Science with Theology in Christendom* (1895), the most visible manifestation of 'the warfare thesis' on relations between science and religion. Against this background, White and Titchener established at Cornell a secular and scientific approach to mind in a department of psychology. Titchener used experimental methods and undertook studies of the elements of consciousness. He translated work from the German and published his own experimental manual and student textbooks. More firmly than anyone else in North America, he promulgated psychology as a disciplined training in natural science.

Titchener appeared to his colleagues to be something like Wundt's representative. In fact, Titchener was not sympathetic with either Wundt's philosophy or his *Völkerpsychologie*, and he brought across the Atlantic a theory of knowledge derived from Locke and undiluted by his residence in Germany. He thought that scientific psychology must start with experiment to refine objective description of the elementary units which combine to construct mental content. While he assumed that the combination of elements reflects causal nervous processes, he believed that psychology, at least for the time being, must concentrate on empirical description. As Titchener defined psychology as the scientific study of the elementary constituents of the conscious mind, he was not interested, in contrast to many of his colleagues, in animals or children or individual differences. Further, he berated his colleagues for rushing to apply psychology before they had established the scientific base of the subject. He maintained a distinctive stance, as he gave priority to training scientists rather than to a psychology concerned with how people function in society. As a result, he was a respected figure within the profession but somewhat marginal to the expansion of psychology as an occupation.

By the time the United States entered the First World War in 1917, psychologists had sufficient confidence in their subject to make a large-scale bid to contribute to the war effort. What they offered was personnel selection – and the next chapter will sketch the background. Their discipline had grown rapidly over the previous 30 years and become a force in higher education and a practical occupation. If sheer scale is an index, psychology was an American discipline. Such gross measures mislead, however, since psychologists were engaged in many different activities and did not agree about the definition of psychology's subject-matter or the relation of knowledge to social affairs. W. B. Pillsbury (1872–1960), who established a psychology laboratory at the University of Michigan in 1897, went so far as to liken early conferences of psychologists to a 'meeting of paranoiacs in a hospital ward'. The Americans, however, soon settled down to more professional, if sometimes still disputatious, relations.

## READING

There has long been a guide to British psychology in L. S. Hearnshaw, *A Short History of British Psychology, 1840–1940* (London, 1964), now supplemented by G. C. Bunn, A. D. Lovie and G. D. Richards, eds, *Psychology in Britain: Historical Essays and Personal Reflections* (Leicester, 2001), A. Collins, 'England', in *The Oxford Handbook of the History of Psychology*, ed. D. B. Baker (New York, 2012), pp. 182–210, and for detail on debate about the nature of psychology, R. Smith, *Free Will and the Human Sciences in Britain, 1870–1910* (London, 2013). A concise, precise history of French psychology is J. Carroy, A. Ohayon and R. Plas, *Histoire de la psychologie en France, xixe-xxe siècles* (Paris, 2006); also J. I. Brooks, III, *The Eclectic Legacy: Academic Philosophy and the Human Sciences in Nineteenth-Century France* (Newark, DE, 1998). Some comparative materials are in W. R. Woodward and M. G. Ash, eds, *The Problematic Science: Psychology in Nineteenth-Century Thought* (New York, 1982), and more, including Russia, in G. Cimino and R. Plas, eds, 'The Rise of "Scientific" Psychology within the Cultural, Social and Institutional Contexts of European and Extra-European Countries between the 19th and 20th Centuries', special issue, *Physis: Rivista internazionale di storia della scienza*, XLIII/1–2 (2006).

On the Netherlands: T. Dehue, *Changing the Rules: Psychology in the Netherlands, 1900–1985* (Cambridge, 1995). H. Misiak and V. M. Staudt, *Catholics in Psychology: A Historical Survey* (New York, 1954) remains informative. There is a general overview in M. G. Ash, 'Psychology', in *The Cambridge History of Science, Volume 7: The Modern Social Sciences*, ed. T. M. Porter and D. Ross (Cambridge, 2003), pp. 251–74.

The literature by and for psychologists is large for the second half of the nineteenth century, reflecting the standard story of the rise of scientific psychology in Germany and its transfer to the U.S. For informed sources on Wundt: R. W. Rieber and D. K. Robinson, eds, in collaboration with A. L. Blumenthal and K. Danziger, *Wilhelm Wundt in History: The Making of a Scientific Psychology* (New York, 2001); also G. Hatfield, 'Wundt and Psychology as Science: Disciplinary Transformations', *Perspectives on Science*, V (1997), pp. 349–82. On Helmholtz, G. Hatfield, *The Natural and the Normative: Theories of Spatial Perception from Kant to Helmholtz* (Cambridge, MA, 1990), and R. S. Turner, *In the Eye's Mind: Vision and the Helmholtz–Hering Controversy* (Princeton, NJ, 1994). For Fechner, M. Heidelberger, *Nature from Within: Gustav Theodore Fechner and His Psychophysical World View* (Pittsburg, 2004). From a sociological perspective, there is detailed work by M. Kusch: *Psychologism: A Case Study in the Sociology of Philosophical Knowledge* (London, 1995). For an introduction to Dilthey, R. L. Anderson, 'The Debate over the *Geisteswissenschaften* in German Philosophy', in *The Cambridge History of Philosophy, 1870–1945*, ed. T. Baldwin (Cambridge, 2003), pp. 221–34.

To my knowledge, the best history of the rise of experimentalism in the U.S., especially in the educational context, is J. M. O'Donnell, *The Origins of Behaviorism: American Psychology, 1870–1920* (New York, 1986). For psychology as moral project: C. J. Karier, *Scientists of the Mind: Intellectual Founders of Modern Psychology* (Urbana, IL, 1986); G. Richards, '"To Know Our Fellow Men to Do Them Good": American Psychology's Enduring Moral Project', *History of the Human Sciences*, VIII/3 (1995), pp. 1–24. R. E. Fancher, *Pioneers of Psychology* (New York, 1996), is readable and informed.

CHAPTER FOUR

# Psychological Society

*What are we?* Psychology is the only means by which this momentous question can be fully answered.
C. K. Ogden

## WHO IS A PSYCHOLOGIST?

During the twentieth century, psychology became a cluster of specialities, each with large-scale investment, a distinct agenda and an institutional identity. After the Second World War and the collapse of humane values, there was a sharp incentive for rapid growth in the sheer quantity of psychological activity. No one can do justice to this by citing a few influential authors, intellectual movements and scientific schools. There were hundreds of psychologists who hardly left the laboratory or lecture hall, but there were thousands who worked in commercial business, the military, hospitals and schools. Psychological ways of thought also flourished in the everyday life of ordinary people, and Mr and Ms Everyman became psychologists themselves. Psychological knowledge acquired a taken-for-granted quality, and this altered subjectivity and changed expectations about what it is to be human. There was a new degree of emphasis on the personal world in psychological terms, on the psychological self, with ramifications in every aspect of Western life.

I therefore write about *psychological society*: society in which the psychological representation of life became dominant, the norm. The phrase 'psychological society' appears to have originated or spread with a book, given this title, by the American medical journalist M. L. Gross in 1978. Gross used the phrase to characterize the uptake of his particular *bête noir*, psychotherapy, and he did not at all indict other approaches, notably biological ones, which similarly emphasized the individual and psychological route to understanding people and their problems; indeed, he barely talked about society at all. When I take up his phrase, I am referring to something larger, which certainly encompasses biological forms of psychology as well, and highlights the fact that modern individualism, expressed in so much psychological belief and activity, is a type of *social* order.

The nineteenth century was an age of activism and reform through educational innovation, new institutions of asylum or punishment, legislation about health and work and much else besides. Here, it has been suggested, rather than in academic pursuits, are the roots of psychology and the social sciences generally, in the administrative and institutional means to manage people or to teach people to manage themselves. Schools, prisons, asylums, institutions for self-improvement, workhouses, families, government reports and bureaucracies, charities, church groups, youth movements and factories turned human beings into systematic objects of study. These public sites, the argument proceeds, created the core of the subject-matter of specialized knowledge in modern psychology – the people identified as 'a problem': school children, criminals, people with mental disabilities, and also those who grieve, are lazy, who divorce, who take drink or drugs, or generally are just too wild or too passive. Late eighteenth-century German teachers kept diaries in order to understand their pupils' development; the managers of Auburn prison in New York in the 1820s made biographical sketches of their inmates and tried to understand the origins of crime. The activity of psychologists originated with the identification, or even the coming into existence, of groups of people who had a psychological problem. Institutions like school or prison placed people in a circumscribed environment in which they became objects of study. Those who carried out the studies generalized from these environments to truths for which they claimed scientific authority and which they held applied to the population at large. The social institutions were thus, in effect, early laboratories; but the people who worked in them were not, at least initially, specialists.

There has been a circle of influences. People's lives gave psychology its subject-matter, and knowledge about this subject-matter suggested changes in how to live. Schooling in mass society created pressure to differentiate individual capacities, and the knowledge which resulted – IQ scores – informed people's self-image as well as educational policy. This is psychological society: the mode of life has psychological form, and it is a secondary matter, for this description, whether its language is psychoanalytic, phenomenological, biological or whatever.

A number of organizations existed whose members called themselves 'psychologists' and claimed that training, expertise and scientific objectivity distinguished them from non-professionals. The early British Psychological Society, for example, restricted membership to 'only those who are recognised teachers in some branch of psychology or who have published work of recognisable value'. The initial membership was tiny, and the preponderant training was, in fact, in physiology. Nevertheless, as I have commented earlier, psychology was from the beginning a public pursuit. There was an inherent difficulty in saying who was or was not a psychologist, and, as psychology spread, the difficulty was compounded. People are psychology's subject-matter, but people are neither passive nor stupid and take this subject-matter, themselves, to be a matter for themselves. Common language differentiated 'scientific' psychology and 'popular' psychology, but there was no absolute line.

Of course, there were, and are, occupational differences between psychologists in academic settings and in social administration or commerce, as well as large differences between trained psychologists and 'pop' psychologists. Since academic psychology developed on a huge scale, especially after 1945, with its own rules and practices, at times it was to a considerable extent autonomous from ordinary life. Running rats in mazes to test a theory of learning, for example, was remote from teaching children to learn. Nonetheless, the line between scientific and popular psychology was never simply given. It is true that trained psychologists, following the pattern of doctors, to whom the state gave special rights and obligations to practise medicine, have tried to police who calls herself or himself a psychologist and who, on this basis, can intervene in people's lives. Some jurisdictions passed legislation about this in the twentieth century, though others not, and it remains a live concern, the history of which varies from country to country. In France, for instance, there has been a particularly tortuous struggle over whether psychoanalysis (even which form of psychoanalysis) is a psychology with scientific standing. In Russia after the Soviet period, the number of people offering some kind of counselling or psychotherapy grew apace, disturbing psychologists with a scientific background. However, there was no effective control. Beyond the details, the point is that struggle over who may legitimately 'do' psychology is endemic because of the significant sense in which ordinary people have learned to be psychologists able and willing to describe life in psychological terms.

Twentieth-century Western societies were certainly psychological in their language about individual capacities and differences, in their approach to identity and in their thought about social policy. From the cradle to the White House, people acquired and expressed identity in terms of a psychology of *difference*. They learned to emphasize psychological separateness, individuality, the language of the inner self and of outward capacity. In mundane but pervasive ways, belief in individual psychological selves became of a piece with belief in the individual as the economic, consuming, moral and political actor in liberal democracies. In the twentieth century, again especially in the post-1945 settlement, people internalized belief that each person has the right and duty to make choices, founded on individual differences of inward feeling and outward capacity, as the basis of social life. In the light of this, they looked to psychology to describe identity, to explain choice and to help to act with more freedom.

Psychology, I suggest, flourished – and flourishes – because it was the means to square a circle at the heart of modernity: it appeared to meld, humanely and democratically, individuality, the values of liberal politics and effective social management, reconciling the interests of the individual and of society. There has been a rationalization of life, in the sociologist Max Weber's celebrated phrase, bureaucratic regulation and legally enforceable conformity, to make possible vastly complex social relations. In this life, psychological practices sustained a focus on the individual and treated her or his capacities as real. Further, since people, thinking psychologically, conceived of agency as theirs and feelings as theirs, they recreated the external social order as an internal subjective world: they internalized the social. When this failed, when, for example, a person became so depressed it disrupted work, or when someone, like a young child, did not have the required capacity, psychologists were ready to act both for the client and on behalf of society.

Psychological descriptions of difference never were, and never are, neutral. They arose with management of good and bad character in the nineteenth century and earlier. Study of difference fostered both clinical methods, differentiating the symptomatology of normal and abnormal, and statistical techniques, defining the average and deviation from it. Descriptions of individual human nature presupposed a standard, a *norm*, against which to assess variety, or deviation, in all its forms, and as psychology developed it assembled the techniques with which to determine this norm. The history of the notion of intelligence is central to the story. Moreover, when nineteenth-century psychologists promoted the study of individual capacities, they also promoted a society, a meritocracy, in which order and place depended on individual capacities and not, for instance, on God, inherited status or abstract ideals. Even when psychologists did not acquire authority, as in fact they often did not, since political life is generally opportunistic, there was a groundswell of rational, reform-minded opinion which thought they should have.

## INDIVIDUAL INTELLIGENCE

Everyday arts for assessing character, aiding individual decisions and guiding the passions were ancient and widely familiar. Medicine generally spread a language about individual character or temperament and inherited disposition (the 'diathesis'), and a medical interest in why some people and not others have health or suffer illnesses and recover or die strongly influenced early individual psychology. There were deep roots in humoral medicine and in ethics requiring physicians to respond to the individual patient. A huge religious and moralistic literature discussed character and the fate of the individual soul. Phrenology continued in the United States and Britain in the second half of the nineteenth century, especially in the hands of the entrepreneurial Fowler family who carried on a business offering advice to clients. The Fowlers marketed the still familiar china busts, with the head marked up with the location of the various mental faculties. All this created an audience receptive to the claims of the new expert psychology in education and mental testing.

The inspirational figure for much of the modern interest in individual differences was Francis Galton. Motivated in the 1860s by Darwin's theory and by his own fears about 'the condition of England', that is, the people's persistent poverty, Galton became fascinated by individual differences, heredity and their consequences for society. His creative intellectual step was to argue that the way to study individual inheritance is to study the distribution of variation in populations, and not directly to study the physiology of heredity (which was indeed then fraught with difficulty). The data of statistics also strongly encouraged people to think of human beings, like the rest of nature, as subject to natural law. If an individual event, like an accident, appears a matter of chance, the measurable, regular occurrence of the event in a population suggests the existence of underlying causal laws.

Galton had a sacred respect for the laws of nature. To bring knowledge of people within those laws, he was convinced, requires recognition, first, that mental variation is inherited in the same way as physical variation, and second, that this natural, inherited variation is overwhelmingly responsible for character. He had no time for the soul (though he may have been sensible to a distant cosmic purpose), and he specifically opposed Victorian belief in the moral will, self-help, determining a person's life (though he was also a moralist). He set belief in 'nature' and 'nurture' – these were his terms – against each other and threw the weight of his research behind the former. Convinced about the continuity of biological evolution and human advancement, he concluded that progress depends on the quality and distribution of inheritable variations passed down from generation to generation. The debate, the methods and the prejudices which Galton articulated still haunt the field.

Galton set out to show that heredity rather than environment dominates individual differences, beginning with *Hereditary Genius: An Inquiry into Its Laws and Consequences* (1869), a book full of biographical data supposedly vindicating belief in heredity as the principal cause of individual mental as well as bodily achievement. He devised ways to describe and measure variation in the first place; for example, he set up an anthropometric (or human measuring) laboratory at the International Health Exhibition in London in 1884, where members of the public tested themselves for such things as muscular strength and visual acuity. As a number of other people were doing at the time, Galton studied the variation of fingerprints and introduced a workable classification which led to their use in court, initially to identify recidivists. In *Inquiries into Human Faculty and Its Development* (1883), he introduced the term 'eugenics'. Believing that industrial society and international competition placed growing demands on individuals, he advocated a social policy to increase the number of people with the amount of energy and intelligence needed. He called for a policy to solve social problems and advance human well-being by the promotion of differential birth rates for groups with different inherited aptitudes. He was especially keen to make it easier for young people of good stock to marry and have many children. In the decade after 1900, with fears of degeneration widespread and intense international competition a reality, it was a programme with appeal.

Though no mathematician, Galton perceived the importance of statistical analysis for studying the distribution of individual variation in populations. Taken up by others, this led to modern quantitative methods in both biology and psychology. Galton himself built on the work of the Belgian astronomer and state administrator, L.-A.-J. Quetelet (1796–1874). Since the late eighteenth century, it had been known that observations of a particular phenomenon, like the time at which a star crosses a hairline in a telescope (referred to in connection with reaction-time studies), or the height of soldiers, were regularly distributed around a mean. Quetelet and others analysed the data as if they were divergences or errors in relation to an ideal. In Quetelet's hands, this resulted in the notion of *l'homme moyen*, a hypothetical average and *normal* person. Here was an apparently objective means for studying normality – and deviation from it.

Quetelet was both the leading light of the Belgian state's statistical service and the founder of international congresses on statistics which set out to consolidate the subject as a scientific discipline. Louis-Adolphe Bertillon (1821–1883) took the work further, and he was appointed to a chair in Paris for his work on demographic questions. With brevity and accuracy, Bertillon declared at a congress of statistics in 1855: 'This modest science . . . has become . . . humanity's *bookkeeper*.' One son, Jacques Bertillon (1851–1922), headed the statistical service and focused concern on the low or even declining birth rate in France. Another son, Alphonse Bertillon (1853–1914), developed a system to use photographs of the frontal and profile views of criminals for identification. His system, '*bertillonnage*', taken up by the judicial administration in Paris in 1882, classified appearance based on the distribution of variables around a certain number of means.

It was a step forward in statistical theory when Galton began to study variation as an object of knowledge in its own right rather than as data ('error') with which to calculate a mean. He showed that variation had a standard distribution, a phenomenon which, in the 1890s, the English applied mathematician Karl Pearson (1857–1936) described formally as the normal distribution (or bell-shaped) curve. Pearson, in 1911, became the first Galton Professor of Eugenics at University College, London. Galton also began to study covariance by what became regression analysis, the determination of whether two or more variables are in fact significantly related. Pearson and others treated these ideas rigorously and founded mathematical statistics in its modern form. This discipline introduced a new kind of technical expertise into the field of psychology, and familiarity with statistics did indeed begin to differentiate the expert from the popular psychologist, as generations of psychology students were painfully to learn. Importantly, Galton's work also pioneered the study of psychology as a quantitative subject with data collected from ordinary people in ordinary life. This was a route to a science of psychology based in the social world, not the university. It made it possible for the scientist to enter society's institutions rather than to require society to enter the laboratory.

Galton, and many psychologists after him, analysed a social phenomenon – the distribution of psychological variation in human populations – as the incidence of variations with individual psychological causes. This was to become the hallmark of psychological society: belief that social policy issues are 'really' dependent on individual, psychological factors rather than on structural features of social arrangements. Galton himself went further because he assumed the correspondence between an individual's *innate* capacities and the social position which she or he occupies.

Indeed, in *Hereditary Genius*, he used the high social position of people in successive generations of the same families as prime evidence for the inheritance of high ability. He thought British society relatively open and a society in which natural talent found its natural social place; for reformers, this was a staggeringly false view. Moreover, Galton never discussed women's capacities as a distinct topic. He was conservative and elitist in the proper sense of these words, as he believed the existing social hierarchy was a natural and desirable result of the distribution of individual talent. Attempts to alter it, vociferously expressed by socialists and feminists in the 1890s, could, he thought it objectively possible to prove, only damage the nation. He believed in meritocracy and thought that was precisely what Britain had.

It was Charles Spearman (1863–1945) who developed the technical side of Galton's psychology. Spearman began a military career but switched to science and studied with Wundt in Leipzig as well as elsewhere in Germany before joining the psychology department at University College, London, in 1906. While in Germany, he published the paper 'General Intelligence Objectively Determined' (1904), which guided all his future work and, after he became head of his department, set much of the agenda for psychology in Britain. Familiar with the contemporary work of Binet, he postulated that a general factor (g) underlies all mental performance or intelligence. Special factors (s) affect particular kinds of performance, for example in music, but Spearman and those who followed him argued that there is a correlation of ability across a broad front, such as between a child's abilities in music and in mathematics. This correlation reveals the g factor. Spearman assigned psychology the task of describing human life in terms of the 'two-factor theory of intelligence', measured as general and special ability, summarizing the argument in *The Abilities of Man: Their Nature and Measurement* (1927). The public tended to refer simply to level of intelligence as a general guide to a person's ability and to intelligence as the *cause* of ability.

References in a modern sounding way to intelligence, and to absence of intelligence, had existed from the seventeenth century, but the word was particularly attractive to psychologists in the late nineteenth century as a way to denote a scientific rather than philosophical conception of reason. Philosophers from Plato to Hegel had taken reason as their subject and concentrated on it as a logical procedure rather than as a natural process. Calvinist theologians, concerned with individuals who apparently by nature could not understand Grace, followed by Locke and eighteenth-century philosophers began to change this, thinking of reason as natural activity of mind. It was evolutionary theory, however, which represented mind in fully naturalistic terms., After Spencer, references to intelligence commonly replaced references to reason to denote a capacity of animal as well as human minds. This usage featured especially in a series of books by Romanes amplifying Darwin's evidence that there are no differences in kind, however large the differences of degree, between human and animal mental abilities. In *Animal Intelligence* (1882), written for a general audience, he referred to intelligence as the capacity to adapt to changing circumstances, a definition which he suggested to Darwin and which the latter employed in his studies of the 'intelligent' earthworm's capacity to respond to leaves blocking the entrance to its burrow. If reason had once been thought unique to human beings, intelligence united humans with the animals.

There was lasting debate about the notion of general intelligence and Spearman's claim that it depended on underlying biological rather than environmental causes. Divergence of opinion in the field of factor analysis, though, did not prevent social policies being adopted which treated mental factors as real, natural objects. The reduction of intelligence to a single measurable factor had political consequences. A British government report of 1938 stated: general intelligence 'appears to enter into everything which the child attempts to think, to say, or do [. . . and it is therefore] possible at a very early age to predict with some degree of accuracy the ultimate level of a child's intellectual powers'. The 11-plus examination system, directing children to different grammar or secondary modern schools, with very different academic prospects, subsequently put this assumption about predictability into effect.

Much ingenuity and sophisticated mathematics went into the development of psychometric methodology and the analysis of results. This was sometimes at the expense of the truism that no amount of arithmetic subtlety can replace the need for appropriate measurements across an appropriate range of data in the first place. As the wise woman says: rubbish in, rubbish out. What, then, were psychologists interested in when they observed intelligence?

This question points again to the importance of education in psychology's history. Given the large expansion of public education in many countries in the late nineteenth century, there were, naturally enough, calls to make education a science, and this in turn called for a psychology of the child. In the 1880s and 1890s, for example, there was a vogue in the English-speaking world for Herbartian education, named after the Prussian philosopher and pedagogue, which stressed that the child's, not the adult's, interest, developing in stages, should be the driving force in teaching. The reality of mass schooling, however, was all too often large classes, a standard curriculum and instruction indistinguishable from the discipline which was its precondition. Educators referred to intelligence in comparing a child's performance relative to that of other children and to the expected standard, measured by examinations. It was a small step to refer to intelligence as if it were a natural object, a 'thing' in each child, and thus to make intelligence the explanation rather than the measure of performance. The social nature of the measurement process disappeared from view, while the psychological factor supposedly observed gained prominence. This was characteristic of psychological society.

The architects of the French education system wanted a selection procedure for the best students, but they found that it was pupils with lower rather than higher abilities who placed greatest demands on smooth administration. During the 1890s, Binet began to get involved, and the outcome was the first system of mental testing in order to measure mental difference. Binet was a wealthy young man who found his vocation in psychology. After an initial interest in hypnotism, he turned from medical diagnosis to psychological study as a more reliable way to describe individual capacities. Binet had found marked differences in individual responses to hypnotism, and he carried this interest into new work. He made imaginative studies of his own two children, Madeleine and Alice, noting, for example, stylistic and temperamental differences in the way they learned. He also observed differences in manner between a child's and an adult's intellect. Then, in 1891, he started working – unpaid – at the new laboratory of physiological psychology at the Sorbonne, and in 1894 he became its director after the retirement of its founder, Henry Beaunis (1830–1921). He then actively promoted an experimental psychology informed by interest in individual differences.

Though Binet and his colleagues and students sometimes drew on techniques to measure sensory discrimination and reaction-times developed in Germany, their approach was distinctive in two respects. First, they favoured in-depth studies, analogous to clinical examinations, of people special in some way. Binet himself published studies of *grands calculateurs* (exceptionally gifted calculators) and of people with a phenomenal capacity at chess (who were, for example, able while blindfolded to win several games simultaneously). Second, their research often attempted to compare high-level mental capacities, such as comprehension or aesthetic appreciation, in contrast with German research on the elements of perception. In effect, they examined the capacities which matter in the social world as opposed to the artificial world of the laboratory. Binet, with Victor Henri

(1872–1940), published an overview, 'La psychologie individuelle' (1895; 'Individual Psychology'), which gave this field a recognizable identity and proposed a series of topics for research. In the psychologists' view, the driving force of the subject is 'to illuminate the practical importance that [the topic has] for the pedagogue, the doctor, the anthropologist and even the judge'. Management of children, the mentally disordered, differences of race and criminals gave psychology both its subject-matter and its social significance. And so it continued for the next century.

In 1904, the French government established a commission, to which Binet was appointed, to investigate the mentally 'subnormal' in schools. (Any word I use is loaded.) It was difficult to diagnose slight subnormality; individual children were diagnosed differently by different criteria, borderline children often wasted much of a teacher's time and there was the question of the stigmatization of children who were perhaps not really subnormal. The administrative goal was clear: to identify subnormal pupils early and accurately and to take them out of the standard classes, allowing the latter to make normal progress and the abnormal pupils to receive special education. But subnormality had significance beyond the classroom. The subnormal person was at the centre of contemporary fears about degeneration in the population, especially in the urban working class. The evidence for degeneration was low reproductive rates and high levels of alcoholism, insanity, crime and prostitution. These were the fears which encouraged support for eugenics. The whole area was already a large interest of doctors, and the new psychological researches impinged on a medical preserve. The mental test, of the kind pioneered by Binet, became the tool which legitimated the psychologists' claim to independent expertise and professional recognition. This was to be pivotal in the development of an occupational field of clinical psychology.

Binet, working with Théodore Simon (1873–1961), a colleague with experience of an institution for retarded children, searched for tasks which would reliably discriminate normal and subnormal children. They succeeded when they realized that the tasks had to take age into account: subnormal and normal children might accomplish the same tasks, but the former did so at an older age. They therefore set up a standard, published in 1905, of what a normal child at a particular age can do based on a series of 30 tasks of graded difficulty. Though the tests were designed for a practical purpose, they examined a variety of relatively complex mental functions and therefore appeared to assess a general psychological capacity (which Binet and Simon called 'judgment') of theoretical interest. Since Spearman, in the previous year, had used correlation coefficients between different mental abilities to reveal general intelligence, psychologists immediately set to work on a general theory of intelligence.

Binet and Simon revised their test in 1908 and again in 1911 (the year in which Binet died) in the light of wider experience with larger numbers of children. The uptake of the rather inflexible tests was slow in France, though they were adopted in Belgium and explored by Édouard Claparède (1873–1940), who founded and headed the Jean-Jacques Rousseau institute in Geneva. In Paris, Henri Piéron (1881–1964) continued Binet's work and, taking over the Sorbonne laboratory, created an institute and, in effect, the first French graduate school in psychology. Piéron, who had broad interests, made it possible for French psychological research to expand and flourish both in the academic world and in the applied setting. With his wife, Marguerite, who did the testing while he did the writing, he perpetuated Binet's style of description, which drew up individual profiles rather than constructed a theory of general intelligence. The Piérons' methods were then used, for example, to screen applicants for work on the Paris transport system and on the railways.

The versatile German psychologist, Stern, also made a significant contribution. Stern objected in general to psychology which failed to centre itself on the individual, and he accused fellow German psychologists of being more concerned with abstract notions of mind than real people. Thus, like Binet, he was interested in individual differences and in turning their study into rigorous science. His work also included child development and what he called 'applied psychology', for which he organized an institute in Berlin. Bringing all this together in *Die psychologische Methoden der Intelligenz-prüfung* (1912; *The Psychological Methods of Intelligence Testing*), Stern refined the index of a child's intelligence derived from the Binet–Simon test by dividing the mental-age score by chronological age. He called this index the 'intelligence quotient' or 'IQ'. Devised as convenient language for the comparison of test results, it was easy for people to refer to IQ as if it described a real, quantifiable entity in human nature, and then to suppose that this entity is part of genetic inheritance. This was a common pattern: descriptive language reappeared in psychology as names for hypothetical causal entities. Stern himself did not take these steps. He thought, for example, that when two people show the same level of performance, they may achieve it in different ways, not because they share an inherited cause. He took seriously his own stress on individuality: 'There never is a real phenomenological equivalence between the intelligence of two persons.' In other words, Stern believed, intelligence has different nature and meaning in the context of different people's subjective lives. He elaborated his stance into 'personalism', on which he lectured to diverse audiences as far apart as America and Bulgaria.

Binet and Simon's work created interest among American psychologists and doctors who were predisposed to look for heredity's contribution to social problems. Henry Herbert Goddard (1866–1857), a teacher who retrained as a psychologist under Hall, in 1906 took up what was then a most unusual position as research psychologist at the New Jersey Training School for Feebleminded Boys and Girls in Vineland. He initially brought in the experimental apparatus of an academic psychologist, but he quickly realized that what the institution needed was a reliable means to classify the children in its charge. Institutions like Vineland, which were headed by physicians, housed children with a variety of disorders, later distinguished as epilepsy, autism, behavioural problems and so on, not just 'feeblemindedness'. On a study visit to Belgium, Goddard learned about Binet and Simon's test. He tried it out at Vineland and found that he could achieve results comparable to the assessments made by the institution's staff. Tests did indeed meet a need, as became clear when the American Association for the Study of the Feebleminded, though dominated by physicians, in 1910 adopted psychological tests as the main technique for the diagnosis of subnormality.

Goddard's lasting contribution was to create a home for psychology in medicine. He became better known, however, for a grossly simplified hereditarian argument, in which he attributed human capacities to a unitary intelligence and attributed intelligence to a single genetic factor. In his attention-grabbing book, *The Kallikak Family: A Study in the Heredity of Feeble-Mindedness* (1912), he described two branches of a family, one with a supposed gene for feeblemindedness and the other without. He combined the Greek words '*kalos*' (good) and '*kakos*' (bad) to coin the name 'Kallikak' and told a story in which the family displayed decent normality on one side and an abundance of idiocy and vice on the other. Goddard did not even pretend this was science, but he contributed to a climate of opinion believing that science, and its new instrument the intelligence test, supported hereditarian arguments and social policies based on them.

At a new private university in Stanford in California, Lewis Terman (1877–1956) led a group of educational psychologists who began to apply the tests to large numbers of normal rather than subnormal children. This enabled them to standardize the grading scales, and the outcome was the Stanford–Binet test, the mainstay of intelligence testing in subsequent years. Terman anticipated that objective testing would answer two outstanding political questions: 'Is the place of so-called lower classes in the social and industrial scale the result of their inferior native endowment, or is their apparent inferiority merely a result of their inferior home and school training? [. . . And] are the inferior races really inferior, or are they merely unfortunate in their lack of opportunity to learn?' Whether or not the tests confirmed the natural inferiority of individuals and groups became a hot issue; Terman's gratuitous use of 'merely' reveals were he stood.

A number of psychologists thought that the entrance of the United States into the World War, with the conscription of huge numbers of citizens, offered special opportunities. Conscription brought into the army men of unknown capacity for soldiering, and it was feared that much money and effort might be wasted on those who would turn out to be mentally incapable or unfit for training. Psychologists, led by Robert M. Yerkes (1876–1956), a Harvard professor and president of the American Psychological Association in 1917, worked through the National Research Council to propose to the army a system of mass testing. One and three quarter million men were tested, and psychologists thereby gained considerable managerial experience and a large body of data to use for the comparative analysis of intellectual capacity.

The nature of tests also changed: Binet took a couple of hours to test one child, while the army required large numbers to be tested at the same time. The innovation was the multiple-choice answer technique, which required answers in conformity with predetermined right or wrong choices and did not offer scope for individual people to exhibit capacities on their own terms (as Stern argued should be the case). By such means, psychologists introduced into people's lives technologies shaping self-understanding of what it is natural to perceive, feel and think. Through mass-testing, psychology also gained an audience familiar with and willing to accept its procedures. This occurred despite the fact that the tests used remained somewhat controversial in the academic literature and despite the lack of evidence that the army gained much in exchange for the opportunity given to psychologists. Where psychology did make a mark was in government circles. The appointment in 1919 of the psychologist J. R. Angell as chairman of the federal organization of natural scientists, the National Research Council, was a sign of psychology's coming of age as a reputable natural science. Angell went on to become president of Yale University and to preside over a major attempt to unify the human sciences which took place in the Yale Institute of Human Relations.

Psychologists by and large interpreted test results as support for what they already believed they knew: white home-grown citizens perform better than immigrants; the performance of immigrants decreases the further south and east in Europe their origins; and black citizens are consistently poorer performers than white. Yerkes, Terman and others argued that innate mental differences between peoples are real. Other interpretations of the data were possible and expressed, even early on; it was, after all, scarcely credible that the average mental age of army recruits should be thirteen, as first uncorrected calculations indicated. During the interwar years, a number of psychologists, some from minority ethnic backgrounds, emphasized the role of cultural environment and social opportunity in determining test outcomes. Was it true, then, as Boring remarked in 1923, that 'intelligence is what the tests test'? Faced by the complexities of the mental, behavioural and social worlds, measurement and analysis of the correlations between different factors appeared to be the means to reveal underlying patterns. But did the results reveal causal relations, as those who linked intelligence to heredity thought, or were the results merely descriptive? This was a philosophical and not only a psychological question, since it was debated in general whether science establishes causal knowledge of what is real, or whether knowledge is instrumental, the means to order and hence control the world as the scientist perceives it.

The British debate about Spearman's g factor exemplified this debate. Psychologists did not resolve such questions, but they certainly found occupation in arguing about them. The successor to Spearman's professorship in 1932 was Cyril Burt (1883–1971), who extended the range of factorial analysis as well as its mathematical sophistication. Burt had been appointed the first educational psychologist for the London County Council, the body administratively responsible for the education of London's children. He was therefore at the hub of administrative machinery set up to solve social problems by education, which was a sign that society at large had begun to look to experts to carry through its policies. In this respect, Burt's situation can be compared with Binet's. Burt also acquired an interest in feeble-mindedness and developed tests relevant to educational problems. As he also did clinical work, he never conceived of tests as mechanical exercises. Burt, and the many people active in local educational administration and child welfare with whom he worked, constructed a psychological language for abnormality and delinquency, as Burt showed in his best known book, *The Young Delinquent* (1925). Beyond all this, as Spearman and other psychologists had begun to do, he applied the techniques developed for the study of intelligence to emotional and active life, that is, to moral character and personality. Burt hoped by this means to develop differential psychology (the study of individual differences) into a general psychology.

In the light of Burt's later reputation, it is worth pausing to note the extent to which he immersed himself in the conditions in which his subjects lived. He saw no contradiction between a policy to improve educational opportunity and belief in the determination of mental capacities by biological causes. What Burt failed to do adequately was to articulate the *social* interaction between inherited structure and environment during a child's growth; he did not, for instance, question whether it is possible to study human heredity by the same methods as animal heredity.

## PERSONALITY AND WORK

When psychologists introduced tests it was immediately apparent to them that much more than a measure of intelligence was needed. In advertising, business and the army, as in ordinary life, clients were at least as much interested in energy, application, persistence, desire to get on and so forth. The Victorians subsumed such things under 'character'. The word persisted, but to psychologists it increasingly smacked of moral rather than expert language. Two words then became current in the English-language psychological literature, 'motivation' and 'personality', to describe the causal drive to do things and the qualities of self which develop as a result of doing things one way rather than another. The development of techniques to study and measure motivation and personality became central to psychology in the commercial sphere. Psychologists did not apply knowledge of motivation and personality; rather, they elaborated these concepts as they worked to show their value as experts.

Independently of intelligence testing, U.S. psychologists like Edward L. Thorndike (1874–1949) and Robert S. Woodworth (1869–1962) transformed German experimental techniques in order to record individual differences in the laboratory. They initiated measures of performance relevant and marketable in the world of commerce (advertising), industry (personnel selection), education (streaming) and social administration (delinquency). Thorndike, who had a strong institutional base at Columbia Teachers College in New York, directed an army of educational efficiency experts whom he inculcated with his belief that scientific understanding is equivalent to effective measurement. He rode a wave of optimism about psychology's potential contribution to human welfare, leading to a startling claim: 'History records no career, war or revolution that can compare in significance with the fact that the correlation between intellect and morality is approximately .3, a fact to which perhaps a fourth of the world's progress is due.' Psychological testing, it would seem, was the means to put the scientific outlook at the heart of modern enterprise. Woodworth, also at Columbia, concentrated on motivation (developing a field which he called 'dynamic psychology') and introduced the notion of drive into the literature. He and others devised tests to measure drive, and, as with intelligence, a term with descriptive value in the context of measurement procedures reappeared as a hypothetical factor causing activity. The engineering word 'drive' proved to be irresistibly attractive as a unifying term in studies of motivation: it rendered motivation measurable, which was what interested commercial and administrative clients and conferred scientific standing on psychologists.

The interest in individual characteristics besides intelligence led to a growing area of psychology in the interwar years concerned with personality. The term 'personality' became ubiquitous in the English language of both psychologists and the general public in the twentieth century, and this was a clear sign of psychological society. Earlier, the French word '*personnalité*' was much more common than its English equivalent; Maine de Biran, for example, had referred to '*la personnalité ou le sentiment du moi*' ('the personality or the feeling of self'). It was associated with the subjective sense of self, sometimes but not necessarily tied to belief in the soul, before new psychologists like Ribot and Janet appropriated the term in their discussions of abnormal individuality. Ribot published *Les Maladies de la personnalité* (1895; *The Diseases of the Personality*), and Janet became famous through his studies of the splitting of personality. From the beginning, then, personality, understood as the ensemble of what makes a person an individual, was central in French psychology. The medical case study, providing a unified description of the bodily and mental character of a patient, consolidated the notion of personality as a natural and specific thing.

In English, there was theological precedent for the word 'personality', in which context it denoted the assumption by God, as Christ, of the form of a human person. This employed the word in a most contentious area of Christian faith, the question of the personhood of Christ. With attempts to modernize belief, by the end of the nineteenth century a number of writers were describing Christ's glory in terms of the humility and love with which he became fully human, and they propagated the image of Christ as the model man, the ideal personality. In English, then, 'personality' could connote a moral and spiritual standard of wholeness, with the implication that true personality lies in imitation of Christ.

At the same time, in Britain as in France, a more secular usage became current. Robert Louis Stevenson's *The Strange Case of Dr Jekyll and Mr Hyde* (1886) described the divided self of the protagonist in terms of 'nature and character', and it made no claim to be a psychological, as opposed to a fictional and an allegorical, story. Researchers into spiritualism and psychical phenomena, however, then used 'personality' to refer to what makes for the wholeness of an individual, the wholeness which believers held persists after bodily death. This work culminated in a book by F.W.H. Myers (1843–1901), on *Human Personality and Its Survival of Bodily Death* (published posthumously in 1903). Myers explained psychical experiences by reference to what he called the 'subliminal self', a region of the mind normally below the threshold of consciousness, which, he held, constitutes our essential nature.

This background makes all the more striking the history of personality as a concept in experimental psychology in the United States. There, in the twentieth century, it became a multivariate descriptor of the emotional and active sides of an individual's psychological life, measuring things like emotionality and dependency. Measuring personality addressed the demands of clients, but at the same time, because of the word's resonance with preciousness and individuality, it appeared to align psychology with humane values. Further, for some researchers, investigating the multiple factors which make up personality appeared the route to a unified theory, and hence true science, of human difference.

The commercial relevance was already evident in the work of one of the pioneers of psychology in advertising, Walter Dill Scott (1869–1955). He moved to the business school in Pittsburgh in the first decade of the century and explored the selection of salesmen. Subsequently, he directed the area of testing which was of most use to the army in the war, personnel selection, and this work pointed to the importance of character or personality. In the 1920s, a number of corporations began to perceive value in personality tests. At the same time, proponents of the mental hygiene movement, who sought to create an organic, smoothly functioning society through the adaptation of individuals to modern conditions, found tests to be a valuable means to pick out children constitutionally in need, they thought, of special direction or institutional provision. This linked up with European interests. In the U.S., large funds became available, notably from the Laura Spelman Rockefeller Memorial, to support research and the application of new psychological techniques. It is worth noting just how much, in the interwar period, funding for biology generally stemmed from a commitment to found the human sciences.

The systematic description of a field of personality research was the achievement of Stern's American student, Gordon W. Allport (1897–1967), in *Personality: A Psychological Interpretation* (1937). Stern, who arrived unhappily in the U.S. as a refugee shortly before his death in 1938, hoped that a field of what he called 'personalistic psychology' would do something to counterbalance over-reliance on intelligence in describing and classifying people. North American personality testing did indeed flower, but it was not the answer Stern had sought. In the 1930s, the word 'trait' spread as part of an attempt to bring rigour to descriptive categories. But the effort to make the study of personality more scientific focused on *measuring* traits rather than on *clarifying assumptions* embedded in the relationship between those who carried out the tests and those whom they tested. The skill of psychologists went into the former, not the latter. Most obviously, testing for personality relied on standards of normality, and these standards reflected the culture of testers rather than of subjects.

In continental Europe, it was a science of characterology rather than of personality which flourished. As its practitioners used qualitative as opposed to quantitative methods, their research was not clearly distinguishable from everyday, indeed political, judgments about human types. Scientists who practised character analysis included the Swiss therapist Jung, who analysed character in the language of archetypes, the Dutch psychologist Heymans, who devised a scheme (the Heymans' cube) to represent the temperaments in three dimensions, and the German psychologist Ludwig Klages (1872–1956), who used handwriting to reveal the expressive inner self in an expertise called graphology. In the 1920s and 1930s, Klages's textbook on handwriting and character went through many editions, making him the best known representative of a psychology of character linking outer conduct to the inner soul. There was a particularly strong Dutch interest in graphology, adopted from Klages, as a technique for judging character. Characterology significantly used language which enabled both an expert to respond to a person's problems and a person to reflect on her or his own condition. Moreover, the language was equally appropriate in a pastoral religious setting and in a therapeutic secular setting. Characterology appealed as a response simultaneously to spiritual and adaptive needs of people at sea in the modern world.

When, in the 1930s, Nazi ideologists took over the language of human differences to describe both races and individuals, they built on the existing literature about character and on the public's ready belief in natural distinctions. A notorious case of continuity between differential psychology and psychology under the Third Reich was that of Erich Rudolf Jaensch (1883–1940). He revised his character theory to conform to Nazi race beliefs and, in 1933, took control of the leading German journal of scientific psychology, *Zeitschrift für Psychologie* (*Journal for Psychology*), at the exact moment when a new civil service law removed people classified as Jews, or married to those classified as Jews, from academic employment.

European research on character had a completely opposite influence in the U.S. in the 1940s through the work of refugees associated with the Frankfurt school of political and social science, attached in New York to the New School for Social Research. These intellectuals, naturally obsessed with explaining the European catastrophe, looked for insight in the relationship between psychological character and political events, and this contributed a qualitative and overtly political dimension to the study of personality. In *The Fear of Freedom* (1941), Erich Fromm (1900–1980), writing for the public, explained the Nazi and fascist states' exploitation of individual, psychological weakness in terms of the fear of accepting final responsibility. Research culminated in *The Authoritarian Personality* (1950), a large-scale multi-authored study into the personality type thought conducive to authoritarian politics. Initiated in 1941 as a response to Nazi power, in the United States the book attracted attention and criticism for its methodology in the study of personality rather than for its contribution to politics. But it also contributed to a considerable body of literature on attitude, conformity, morale and other characteristics of people as political actors.

By the time the U.S. entered the war, after the bombing of Pearl Harbor late in 1941, it already had a homegrown body of theory linking personality to society – the culture and personality school. Research on personality had become central to the field of social psychology, and in this context research techniques differed from those of mainstream psychology and drew instead on the field methods of social anthropology. The focus was the description of character as a function of the way of life. The work gained a large audience in the 1930s through two students of the anthropologist Franz Boas, Margaret Mead (1901–1978) and Ruth Benedict (1887–1948). In her book, *Coming of Age in Samoa* (1928), Mead, who did her work while still in her mid-twenties, described female adolescence in Samoa, and in her book she contrasted the easy and carefree experience of Samoan girls with the hang-ups and anxieties of her American contemporaries. Later scholars were to question the accuracy of her work (as they were to question many reports of 'other' societies), but her points were influential. Criticizing the work of earlier psychologists, notably Hall, Mead argued with her vivid example that culture rather than fixed biological determinants control a child's development. Educationalists found support for believing a child's development to be open-ended, and this added weight to contemporary propaganda for society in general and mothers and teachers in particular to create optimum conditions for the new generation. As Mead wrote in the 1930s: 'human nature is almost unbelievably malleable, responding accurately and contrastingly to contrasting cultural conditions'. Benedict's extremely successful book, *Patterns of Culture* (1934), spread the same message.

Mead continued for many years to relate ethnographic results to North American preoccupations with child care, sex roles and personality. In *Male and Female* (1949), she showed, however, that she had begun to rethink the extreme environmentalism of her earlier work. Long before this, psychologists had extended personality tests in order to find and measure indicators for masculinity and femininity. As early as 1903, the Chicago psychologist Helen Bradford Thompson (1874–1947) had tried, though in her judgment inconclusively, to resolve the question whether there are systematic differences in intelligence and other abilities between men and women. In the 1920s, psychologists constructed tests specifically to discriminate female and male qualities, and they found what they looked for. Mead and argument about the roots of personality thus came on top of an existing debate about the nature and nurture of sexual differences. (The word 'gender' did not then have its modern use.) Critics at the time and later argued that male psychologists like Terman and Yerkes built their prejudices into their test constructions. Those who were, like Mead, trained in anthropology and the social sciences tended to be sceptical of the existence of innate psychological differences. Mead, indeed, concluded that 'the personalities of the two sexes are socially produced'.

Psychologists entered the world of business offering personality tests to help with personnel selection, productive efficiency and marketing. They also promised to help people adjust and become better citizens. Across the different spheres of education, commerce and the family, there was a common language of energy, efficiency, work capacity, attention, fatigue and adjustment. Especially in the United States, evolutionary belief merged with a vision of social progress and provided the intellectual rationale for human adjustment as the natural continuation of biological adaptation. It became a cliché in the interwar years to say that 'a man's whole life consists in a process of adjustment to his environment'.

About 1900 there was a shift, perhaps even a quantum shift, in the scale and nature of business practice. There were many signs: the new department stores, such as Bon Marché in Paris, large-scale employment of women office workers, the production line, the concentration of capital and market share in national and even international corporations. Western industrial society re-organized for mass production and consumption. The jump in scale generated new approaches to management, finance, production, distribution, personnel and marketing. Management itself became a subject of study, and there seemed every reason to turn it into a science, however practically oriented. Economic imperatives put pressure on science to yield up the secrets of the efficient management of people, and of self-management, and psychologists responded with the measurement of ability, motivation and personality. Even the use of tests to stream children and allocate educational resources was a kind of personnel selection, though the goal was social efficiency in general rather than the profit of any one business in particular.

The first researchers on management adopted an approach taken from engineering, not surprisingly since contemporary observers thought technology was driving new enterprises. New technology was certainly conspicuous: telephone and electricity supply systems, for instance, began to enter homes and cover the landscape with wires in the 1880s and 1890s. It was an engineer, Frederick Winslow Taylor (1856–1915), who began as an apprentice machinist, who systematically marshalled the techniques of industrial efficiency and applied them to workers. In *The Principles and Methods of Scientific Management* (1911), Taylor laid out arguments to alter both employers' and employees' conceptions of work, significantly taking for granted the reality of a common interest in the maximization of returns. He ignored social and political questions concerning labour and capital in order to concentrate on increasing the efficiency of individuals. He spread knowledge of techniques to quantify the different elements of the production process, notably time study (which involved the separation of operations into measurable units), and this made it possible to record and hence reward individual performance. The results in practice served management goals and attempted to adjust human capacities to the crude technical demands of machines. This was the world of managing people as immortalized in Fritz Lang's *Metropolis* (1927) and Charlie Chaplin's *Modern Times* (1936), films which used a new technology to exhibit the technology, and art, of life in mass society.

The crude human engineering of Taylor's methods was widely criticized. Psychologists had an opportunity to show that they could do better – in connection with both efficiency and human well-being. They could draw on a number of existing studies of work and fatigue focused on the efficient expenditure of both individual and national energy. Psychologists transformed such interests into psycho-technics. Münsterberg, for example, after his move across the Atlantic, wrote an introductory text, *Psychology and Industrial Efficiency* (1913), which set out the psychologist's alternative to Taylor's engineering ideal of human efficiency. It appeared obvious to psychologists that society needed their skills in experimental technique and in the administration of tests, coupled with knowledge about normal development, perception, learning, motor ability and motivation, in order to mould personnel and workplace or consumers and products into integrated wholes. Much factory production, for example, involved repetitive tasks, and fatigue restricted output. Psychologists therefore asked questions about which movements, with what kind of repetition, produce fatigue, whether breaks in work increase the capacity for work and whether incentives affect tiredness and motivation. The British government sponsored such work during the First World War, when the production rate of munitions in the factories at home proved a limiting factor in battle on the Western Front. Transport systems were noteworthy consumers of the new expertise: the Austrian railways, to test staff, ran a special train equipped with psychotechnicians; in Russia, the people's commissariat of labour established a laboratory in Moscow for psycho-technics in 1920, and there were testing services for transport operatives across the USSR.

Another area where psychologists' skills were obviously relevant was in advertising and marketing. During the war, once again, the Johns Hopkins University psychologist Watson produced a film for the U.S. army to warn recruits about venereal disease. He gave thought, not very successfully it seems, on how to stimulate soldiers to refrain from sex, and also on how to monitor the results of the propaganda. In the 1920s, he went on to become an executive of the New York advertising agency, J. Walter Thompson, an innovator in the industry. By then, the advertising industry had twenty years' experience of work with psychologists.

There were a number of new institutional initiatives, like the Soviet institute of labour, to promote practical psychology. Whereas the Soviets looked to the state, elsewhere psychologists were reliant on private or commercial funds. In the U.S. in 1921, psychologists, led by Cattell, organized the Psychological Corporation as a consultancy. But its principal personnel kept one foot firmly in academia and the corporation proved barely viable commercially. However much talk about practical relevance made psychology's growth possible, the significant expansion of U.S. psychology at this time in fact took place within the universities. Substantial employment opportunities for psychologists without an academic base came during and after the Second World War.

Circumstances were different in Britain, where C. S. Myers founded the National Institute for Industrial Psychology in 1921. Myers was familiar with the work of his former student, the Australian psychologist Bernard

Muscio (1887–1926), who had published lectures on industrial psychology given in Sydney and then worked in England for the Industrial Fatigue Research Board in the First World War. Myers concluded that there was a field of enormous social value to be cultivated, and a small but sufficient number of industrialists, as well as the Carnegie and Rockefeller trusts, agreed with him and provided funds. Influenced by his own involvement with the war effort, Myers believed that national efficiency required the informed management of people, and he argued on behalf of psychologists, rather than engineers, as the source of the necessary expertise. Faced by widespread academic ignorance of psychology, and sometimes actual hostility, Myers relied on private initiative and an institution, which he directed full time from 1922, integrated from the outset with the world of practical affairs. The National Institute provided a range of services, such as personnel selection, production planning and marketing advice, to industry, government departments and educational authorities. It also carried out some research, for example, on fatigue, and supported would-be psychologists when opportunities for academic training, let alone academic employment, in the field were rare. It brought people, like the future professor at the University of Liverpool, historian of psychology and biographer of Burt, Leslie S. Hearnshaw (1907–1991), into psychology. Meanwhile, in the area of medical psychology, the Tavistock Clinic had been founded in 1920 with private funds 'to provide treatment along modern psychological lines' for people with mental problems 'who are unable to afford the specialists' fees'. The clinic, from which a children's department separated in 1926, pioneered psychological responses, geared to individuals, to a wide range of human problems. After 1945, it established influential training programmes for professional groups, such as teachers and doctors, and in doing this wove psychological techniques into the fabric of everyday social and medical organizations.

During the second half of the 1930s, the mobilization of German society for war placed a premium on the efficient use of manpower. Womanpower, at least in theory, was sent home to breed. The *Wehrmacht* (the German army) listened to university psychologists who claimed to be able to select officer material. As a consequence of this, academic psychology gained a professional identity and an institutional autonomy from philosophy which it previously had not had. In the middle of war, in 1941, university psychology for the first time acquired a separate programme of study and became a separately examined subject. Its standing with the army did not last, though, since officers selected themselves at a rapid rate under fire on the Russian Front. These circumstances in which professional psychology grew in Germany left an emotionally and politically divisive legacy. After total defeat in 1945, 'Year Zero', many German academics, psychologists included, distanced themselves from the Third Reich and described the period from 1933 to 1945 as an aberration, a political disaster, with which, they argued, the apolitical ideals of science had had nothing to do. There were personal and occupational incentives to claim complete separation of the objective methods of research from the application of knowledge. Psychologists rapidly acquired an interest in behaviourism, Pavlov and experimentally rigorous forms of psychology in the postwar years, and this enhanced their belief that they had no special Nazi legacy. A later generation of German psychologists, however, questioned the role which psychological expertise had had in the war effort, pointed out the continuity of personnel in psychology before and after 1945 and described the way occupational psychology rather than the imperatives of objectivity or the ideals of pure science had established psychology as an academic discipline.

The Second World War was the scientists' war; it was also the psychologists' war. The governments of the United States, Britain and Germany employed technical and scientific experts on an unprecedented scale. The existence of a clear goal – victory – enabled all parties within each country to agree to the politics of planning and centralized decision-making, and governments turned to scientific management. The same ideals had already informed the declared policies of the USSR. The modern management disciplines of systems analysis and operations research both originated under the exigencies of war. In Britain, the medical director of the Tavistock Clinic, J. R. Rees (1890–1969), headed a comprehensive programme of psychological services in the war, an appointment which represented a coming of age for both psychological techniques of personnel selection and for clinical psychology. National validation of the Tavistock orientation reinforced the culture of psychological society, though there remained ambivalence about the place of psychoanalysis. With the clinic integrated into the new National Health Service, a non-clinical Institute of Human Relations opened in 1948 to do research and give training in group psychology methods to people from industry, voluntary care organizations, social work and other areas. There was also the vast task of organizing whole populations for war itself. In terms of sheer scale, testing and personnel selection (for radar monitors, to take one instance) remained psychologists' most important activity. The U.S. army and navy developed versions of the General Classificatory Test (GCT) to help invest effectively in people of the right ability and aptitude. These tests built on the methods of factorial analysis developed by the Chicago mathematical psychologist L. L. Thurstone (1887–1955), who had earlier analysed human performance in terms of what he called 'primary mental abilities', his substitute for crude notions of intelligence.

In Western Europe and the English-speaking world, industry, government, the welfare and health services and education were eager to deploy psychological expertise. In the immediate postwar years, there was widespread political support for the state taking responsibility for welfare. This overlapped with belief in expertise as a force independent of politics. The 1930s and 1940s taught what fanaticism and ideology achieve, reason suggested that psychology and the social sciences generally could do better and the public was willing to look to science to advance the common good. Psychology expanded rapidly and became a large and autonomous field within universities which were themselves expanding in response to demands for educated labour and increased access to higher education. To this I will return in the concluding chapter.

Humane in intention, psychological interventions in health and in the management of personal relations were neither politically nor morally neutral. The development of human management transformed power and authority in social relations and built new expectations about how people should live into the mundane, day-to-day occupations of the helping professions and thereby into the lives of those helped. If in an earlier age people had related to each other through networks of custom and obligation, defined by social place and legal contract, in the new age each person was to have an internal psychological space in which to achieve adjustment to others, satisfactory performance and moral maturity. Occupational and applied psychology were there to help.

## PSYCHOLOGY OF THE CHILD

In continental Europe, war, revolution and conflict, on a scale foreign to Britain and North America in the twentieth century, gave the quest to comprehend suffering and its causes a special intensity. Men and women turned to examine human nature itself, perhaps to understand how terrible things are possible or, more optimistically, to pull up human nature by the roots and change it. Rationalists believed that if only science could acquire a voice in education and the state, it would achieve the humane future which politics had failed to bring about. Hope rested, above all, with children. The child was the largest factor in the involvement of masses of people, especially women, both trained and untrained, in the project of psychology. But the child was not simply a natural object awaiting observation and study. The child was a projection of aspirations for a better life.

The writings of Locke, Rousseau, Pestalozzi and many others had long given authority for an optimistic view, for belief in the human capacity to remake life through children. Child study and research on development and welfare existed in many countries in the nineteenth century. The New York Children's Aid Society, in 1854, began to collect urchins off the city streets and to put them on trains to work on Midwestern farms. Over the next 60 years, over 120,000 children made the journey, with favourable results as measured by the way the children's subsequent careers were interpreted. This policy was the work of idealists and reformers, and it continued through the years which historians often picture as dominated by hereditarian beliefs. Then, in the twentieth century, it was above all the child who made opportunities for experts and carried expert values into family life.

The history of psychology in Vienna offers a rich case study. In the 1920s, indeed to some extent until the *Anschluss* (the Nazi incorporation of Austria into the German Reich in 1938), psychology, especially the psychology of education, flourished. These were the years in which Freud made Vienna the mecca for psychoanalytic training, the socialist city administration integrated educational and clinical support, and the collective provision for child welfare was perhaps unequalled. Young intellectuals placed their idealism at the service of scientific research and education. The university of Vienna made no formal arrangement for psychology until 1922, when it called Karl Bühler (1879–1963) to a chair and at least nominally equipped him with an institute. There had been philosophy professors with a major interest in psychology, notably Brentano and Mach, and Freud had unpaid professorial status in medicine from 1902, but Bühler contributed a new breadth of experience with German theoretical and experimental psychology.

Earlier, in Würzburg, he had been interested in thought processes, contributed an experimental study of the 'Aha!' experience of sudden insight and written a substantial work on language. A highly cultured scholar, in Vienna he supported a humane, child-oriented but scientific, vision of psychology's tasks. Moreover, his appointment was, in effect, a double one, since his wife, Charlotte Bühler (1893–1974), who had been his student, proved to be a formidably efficient manager of the new institute and co-ordinator of a research programme on child development. A vivacious, some say vain, woman, she provided a scientific base for people who looked to the child and education as the means to create a new world. When she attracted Rockefeller funds in 1926, she turned a child study centre, run under the auspices of the university institute, into a centre unique in

Europe. Karl Bühler wrote influential studies on the child's development of language and a survey, *Die Krise der Psychologie* (1927; *The Crisis of Psychology*), in which he attempted to synthesize competing schools of theoretical psychology.

The Bühlers had rigorous notions about objective research. Karl Bühler was sympathetic with, though not actually a member of, the Wiener Kreis, the Vienna Circle, meeting from 1925 until 1936. This circle of positivist philosophers argued that meaningful propositions are of two kinds, either formal logical axioms or empirical statements verifiable by observation. Karl Bühler wanted his students and assistants to attend meetings of the Kreis and learn to demarcate science from wishful thinking or pseudo-science, and within the institute the Bühlers promoted scientific psychology in its own right. Charlotte Bühler therefore instituted a strict regimen of data collection on children in her centre, the purpose of which was to advance a general theory of developmental stages and not to address particular questions posed by education or child-rearing policy. She required a rota of staff continuously to observe babies and young children from a glass-sided corridor. The children, who were there for observation and not treatment, lived with a minimum of human contact in a sterile environment, which supposedly provided a culture-free baseline for observing development.

All the same, many of the Bühlers' students were schoolteachers or involved in teacher training, experimental schools, child guidance or work with disturbed children, and the Wien Pedagogium, a teacher training institute, hired the Bühlers to teach. There was therefore a complex relationship between science and social values. It was the city which funded a child reception centre for the observation of children's problems, and Charlotte Bühler's data came from a specific group of children who had been brought together as a result of illness, social dislocation and family breakdown. Though the research methods were self-consciously scientific, the subject-matter arose as a response to a social problem. This pattern was not special to Vienna, as the earlier child study movements in a number of countries demonstrate. Charlotte Bühler's own ideas became widely known in the U.S. after her emigration and through her later publications on the life cycle.

There was a high level of political activism among the teachers and observers who worked with the children. Some, like Marie Jahoda (1907–2001), were active in the Austrian Social Democratic Party and were informally Marxist in orientation; others, even if they were not overtly political, were involved with educational programmes to create new, socialized and fulfilled human beings. In 1920, Freud's ex-colleague, Alfred Adler, for instance, began an energetic campaign to support therapeutic education, and he pioneered a consultation service for parents and teachers, new kindergartens and experimental schools. There was faith: if the child's nature were known and its needs understood, then its future could be assured; and the dream was that these new children would remake society as a place fit for humanity. In reality, though, Charlotte Bühler's two paid assistants were relatively conservative in politics, money was desperately short and it was private wealth which funded much of the innovative work.

The idealism is illustrated in the career of Lili Roubiczek Peller (1896–1966), born into a wealthy Prague family and trained as a teacher before she travelled to Vienna to work with Karl Bühler. She strove to bring together psychology and the needs of working-class children. Inspired by Maria Montessori, the Italian educationalist who extended techniques devised to teach children with learning difficulties to all children, she created Der Haus der Kinder (The House of Children) in the slums of Vienna's 10th district, which was the first official Montessori school in Austria. With funds from England arranged by Montessori, Peller joined with five women in their late teens to form a collective, and they then built the school with help from local craftsmen and architects. Peller went on to become a consultant to the city's child welfare department, to manufacture Montessori furniture and literature, to offer a training course taken by some of the Bühlers' university students and to develop links with child psychoanalysts. Peller introduced Montessori to Anna Freud, Freud's daughter and pioneer child analyst, in 1930. After she had herself trained as an analyst, Peller worked at the Jackson nursery in Vienna, an observation centre established by a wealthy American paediatrician in 1937. By this date, however, such institutions were politically suspect, and the nursery therefore had no welfare role and it disguised its 'Jewish' intellectual debt to Anna Freud. As she was married to a Jewish physician, and with both Montessori methods and psychoanalysis excoriated by fascists, Peller then emigrated to the United States.

There were other women who had similar careers. How far any of the new psychological interests directly affected children is hard to judge; the authorities neither expected nor received direct practical advice from the Bühlers's child study centre. If the interests fostered a middle-class culture which valued psychology, the way of life was overtaken by politics which did not just denigrate but sought physically to eliminate it. The Austrian and German virulence towards those labelled 'Jews' caught in its hatred large numbers of the Viennese researchers and Charlotte Bühler herself. Those who opposed so-called Jewish psychology conflated this psychology with concern for children's self-expression and the welfare of the working-class child. Austrian fascism and then National Socialism repudiated psychological society. After 1945, the drive to create such a society came from the English-speaking world, informed, however, by the extensive emigration of European psychologists to Britain and North America. In the Netherlands and Scandinavia, indigenous educational and welfare policies then built many of the values of psychological society into the institutions of the state. High levels of political consensus in these countries created circumstances in which psychology and the social sciences burgeoned, and it proved possible to put into practice policies about which the Viennese researchers had only been able to dream.

In North America, faith in education informed a specialized child study movement led by professional psychologists rather than, as earlier, by teachers. The *Journal of Educational Psychology*, founded in 1910, was more experimentally and statistically oriented than the comparable *Pedagogical Seminary*, founded in 1891 by Hall. Even the titles of the two journals stated the difference. There was Rockefeller money for centres of child research at the universities of Columbia, Yale, Iowa, Minnesota and Toronto before Vienna, and these centres were in communication with each other. As early as 1896, Lightner Witmer (1867–1956) opened a psychological clinic at the University of Pennsylvania in which he examined and treated children with learning problems, and he subsequently began to train psychological experts in children's education. Arnold Gesell (1880–1961), at the Yale Psycho-Clinic, opened in 1911, systematically studied babies and young children within a one-way glass dome, and he used measurement and photography to build up a picture of the stages of normal development. He and others then translated such pictures into guides to what every mother could expect from her child, and this supplied a growing interwar industry in motherhood manuals or, to use a later idiom, parenting guides. The mother was expected to become a psychologist. When she did not, and when the child did not develop normally – physically, mentally and morally – expert psychologists were increasingly available to do the job.

In Britain in the 1920s, Burt taught as a professor at the London Day Training College, later the Institute of Education, and this reinforced the position of psychology in the mainstream of British educational practice. During the 1930s, the child study movement changed from being the interest of amateurs, without formal training, to an occupational movement with psychologically trained personnel, specialized institutions and scientific methods (notably diagnostic tests). The Institute of Education established a department of child development in 1933 and appointed Susan Isaacs (1885–1948), who had among much other activity run a progressive school in Cambridge, to head it. Her own career followed a common pattern, as she switched, at least to a degree, from practical engagement to academic psychology.

When Galton gave shape to debate by using the language of nature and nurture, he understood that his own firm hereditarian opinion was in conflict with the Victorian ethos of self-help. His stance, he thought, was founded on scientific reason as opposed to unscientific moralism. He considered at stake science versus non-science, not only nature versus nurture – a conjunction which was to reappear a century later when a new hereditarianism took the offensive. Hereditarian thought became very widespread, though by no means universal, at the end of the nineteenth century, associated with debates about race, the struggle for empire, the rise of socialism and feminism and fears of degeneration, alcoholism and mass society generally. Then the pendulum began to swing. In the 1920s, there was extensive concern with the environmental or social determinants of human capacities, exemplified by the young teachers who worked in the Vienna slums. In this period also, especially in the United States, the modern terms of the nature–nurture controversy became common currency. An emphasis on social determinants remained dominant till the 1960s. Thereafter, the pendulum swung back and, at the beginning of the twenty-first century, many psychologists, though certainly not all, were fascinated by heredity.

One focus of the U.S. debate in the 1920s was restrictive immigration legislation passed in 1923 in response to fears about the number and quality of new immigrants. Since the beginning of the century, an increasing proportion of people entering the country were from Southern and Eastern Europe, Turkey and the Caucasian region and China and Japan. Earlier immigrants feared these people would not integrate into national life and, indeed, believed they were mentally inferior to people of North West European origin. In this setting, the critical issue for test evidence was whether tests presupposed a certain level of linguistic and cultural familiarity with the tester's own world. Many psychologists quickly realized that they did; but their response was not to scrap testing but to search for the holy grail of the neutral test. Practice, critics thought, was to show that testing *always* worked out to favour those with existing social advantages. It was the social act of testing, which constructed a hierarchy of people, not the content of the test, that finally controlled its political and social meaning.

An emphasis on nurture resonated with one side of American political life: faith in the adaptability of people to new social conditions. There was overwhelming support for public education, and through education support for psychology. The 1920s were also the heyday of Watson's brand of behaviourism, which took to an extreme the argument that human nature, judged by behaviour, results from received stimuli. Watson notoriously claimed that any child, if not physically damaged, can become anything. With his student and second wife, Rosalie Rayner, he published widely, often in the popular press, and at least temporarily their opinions influenced advisory pamphlets published by the government. Even when writers gave more attention to the complex developmental stages of childhood, the environmentalist argument placed a heavy burden on parents and teachers: they had the power to form the child. If it was a burden, however, it also bore the democratic aspiration to make each citizen equal, even when circumstances were not equal at birth. When the Germans and Japanese put into practice racism as a principle of world domination in the 1930s and 1940s, the violent antagonism to democratic principles which this involved only served to reinforce their enemies' belief in nurture.

Dichotomies like 'nature versus nurture' polarized debate and all too easily reduced to argument about whether the chicken or the egg came first. The alternative was to build thought on the complexity of the developmental *process* and to reject abstract generalization about heredity versus environment. Conceptually, there was a model for this in the evolutionary theory of the origin of new species through a process, the action of selection (environment) on varieties (inheritance). In the 1890s, Baldwin wrote on the relation between the inherited pattern of development and the 'inherited' social world in which the development occurs. He stressed the role of imitation, leading the child to carry out movements and acquire habits adapting her or him to the social environment. In Geneva, Claparède, in the French-speaking tradition of interest in the individual case, fostered experimental study of individual children's mental growth. In 1921, Jean Piaget (1896–1980) arrived at his institute and began a series of celebrated studies of how and when children acquire perceptual and conceptual abilities and an understanding of the world, causal relations and the boundaries between self and other. He closely studied a small number of children, in numbers without any statistical relevance, on the basis of which he elaborated a general theory of the stages of development. In Piaget's view, there is a regular sequence of stages in which particular capacities emerge, and the presence of each stage and its appropriate exercise is necessary for a child fully to mature. He attributed the child's developmental stages to a biologically evolved pattern, an inherited structure, which organizes experience and action as the child grows, however much the actual expression of each stage depends on the appropriate environment.

The background to Piaget's interest in intelligence was a religious one, linked to the fate of liberal Protestantism at the time of the First World War. As a young man during the war, Piaget hoped in his thinking to integrate the social calamity, which many people at the time attributed to the spread of secular values, with both Christian belief and commitment to the objectivity of science. His solution, boundlessly ambitious, was to identify human reason and morality with the immanent reason of God in the world. The study of the child's development of reason is therefore, he thought, the objective means by which we can know the nature of reason and morality. The culmination of these arguments, and of the child studies generally, was *Le Jugement moral chez l'enfant* (1932; *The Moral Judgment of the Child*), but none of the religious inspiration appeared in print in this or, indeed, in earlier books. Rather, Piaget's readers found a psychological argument demonstrating that the acquisition of moral belief is a stage in children's growth and, hence, a process of natural and normal socialization. Though much of Piaget's influence in the English-speaking world came later, in the 1950s, he provided grounds for optimism about children's acquisition of values. He persuaded parents and teachers to allow education to follow the child's pattern of development; they should accept the way moral learning takes place through the child's interaction with other children rather than through moral instruction.

Baldwin and Piaget were read closely by a young Soviet psychologist in the 1920s, L. S. Vygotsky. For many people, Vygotsky is much the most interesting psychologist to have emerged in the Soviet years – and I shall discuss his work in that context. His overall project was to show how the biological nature of being human and the historical, social nature of consciousness combine in children's development. His reading was exceptionally broad, encompassing both Marxist theory and Piaget's experimental studies, and he elaborated a critical response to Piaget as a means to express his own thoughts. But his work was little known in the West in the 1930s and had to wait until the late 1960s before it had impact.

Though psychology in the first half of the twentieth century became an academic discipline, uniting under a name a number of increasingly specialist occupations, there was considerable two-way traffic between specialists and the public. Nowhere was this more so than in debates about what makes people, individually and collectively, different in intelligence and personality. Children were naturally at the centre, the site where daily life bound together theory and practice, confronting individual capacity with social possibility. In describing psychological society, I have been at pains not to picture scientific psychology as something apart from social life. There was, all the same, much talk about psychology *as science*, and scientific status was clearly central to the authority which psychological ways of thought acquired. What forms, then, did scientific psychology take?

## READING

M. L. Gross's study was *The Psychological Society: A Critical Analysis of Psychiatry, Psychotherapy, Psychoanalysis, and the Psychological Revolution* (New York, 1978). It was the work of the London sociologist Nikolas Rose which persuaded me to emphasize psychological society in the history of psychology: *The Psychological Complex: Social Regulation and the Psychology of the Individual* (London, 1985); *Governing the Soul: The Shaping of the Private Self* (London, 1999); *Inventing Our Selves: Psychology, Power, and Personhood* (Cambridge, 1998). He and I had both read Michel Foucault, who was so influential in part because he re-thought power in social relations just at the time when many Marxian approaches appeared not viable. Though Foucault's career began in clinical psychology, his writings say little directly about it but instead analyse how modern regimes of truth in general express power in understanding being human. Independently, more in the spirit of sociological critique, Kurt Danziger, in *Constructing the Subject: Historical Origins of Psychological Research* (Cambridge, 1990), described in detail the social process by which psychologists themselves constructed the methods and data of their field; in *Naming the Mind: How Psychology Found Its Language* (London, 1997), he discussed the introduction of concepts of intelligence, motivation and personality, among others. M. Thomson, *Psychological Subjects: Identity, Culture, and Health in Twentieth-Century Britain* (Oxford, 2006) is an innovative study of the public culture of psychology. For studies of work and modernity, see A. Rabinbach, *The Human Motor: Energy, Fatigue, and the Origins of Modernity* (Berkeley, CA, 1992), and D. Ross, 'Introduction: Modernism Reconsidered', in *Modernist Impulses in the Human Sciences, 1870–1930*, ed. D. Ross (Baltimore, ML, 1994), pp. 1–25. For the intellectually and scientifically important development of statistics, T. Porter, *The Rise of Statistical Thinking 1820–1900* (Princeton, NJ, 1986).

The historical literature on individual differences and intelligence is comprehensively reviewed and referenced in J. Carson, *The Measure of Merit: Talents, Intelligence, and Inequality in the French and American*

*Republics, 1750–1940* (Princeton, NJ, 2007), a book which has the added benefit of a comparative perspective. A useful introduction is R. E. Fancher, *The Intelligence Men: Makers of the IQ Controversy* (New York, 1985), while M. M. Sokal, ed., *Psychological Testing and American Society* (New Brunswick, NJ, 1985), and L. Zenderland, *Measuring Minds: Henry Herbert Goddard and the Origins of American Intelligence Testing* (Cambridge, 1998) are especially rich for the U.S. There is a still useful biography of Binet: T. H. Wolf, *Alfred Binet* (Chicago, 1973). For U.S. political culture and the development of occupational psychology, there are the thoughtful essays of J. C. Burnham, *Paths into American Culture: Psychology, Medicine, and Morals* (Philadelphia, 1988), as well as D. S. Napoli, *Architects of Adjustment: The History of the Psychological Profession in the United States* (Port Washington, NY, 1981). For measures of masculinity and femininity, M. Lewin, ed., *In the Shadow of the Past: Psychology Portrays the Sexes. A Social and Intellectual History* (New York, 1984), and J. Morawski, 'The Measurement of Masculinity and Femininity: Engendering Categorical Realities', *Journal of Personality*, LIII (1985), pp. 196–223.

For psychology in the Third Reich, U. Geuter, *The Professionalization of Psychology in Nazi Germany*, trans. R. J. Holmes (Cambridge, 1992) is thorough and persuasive. My example of Vienna was inspired by S. Gardner and G. Stevens, *Red Vienna and the Golden Age of Psychology, 1918–1938* (New York, 1992); but for less breathless detail, M. G. Ash, 'Psychology and the Politics of Interwar Vienna: The Vienna Psychological Institute, 1922–1942', in *Psychology in Twentieth-Century Thought and Society*, ed. M. G. Ash and W. R. Woodward (Cambridge, 1987), pp. 143–64. Piaget's inspiration is the subject of a chapter in the same volume: F. Vidal, 'Jean Piaget and the Liberal Protestant Tradition', pp. 271–94; see also his biography of the early years, *Piaget before Piaget* (Cambridge, MA, 1994). There is a fine study of the Victorian background to child psychology in S. Shuttleworth, *The Mind of the Child: Child Development in Literature, Science, and Medicine, 1840–1900* (Oxford, 2010).

CHAPTER FIVE

# Varieties of Science

> Before progress could be made in astronomy, it had to bury astrology; neurology had to bury phrenology; and chemistry had to bury alchemy. But the social sciences, psychology, sociology, political science and economics, will not bury their 'medicine men.' According to the opinion of many scientific men today, psychology even to exist longer, not to speak of becoming a true natural science, must bury subjective subject matter, introspective method and present terminology.
>
> John B. Watson

## THE SCIENTIFIC IDEAL

It would be an interesting project to make an encyclopaedia from definitions of psychology. For J. S. Mill in 1859, it was simply 'the scientific study of mind'; also for James, 'Psychology is the Science of mental Life.' Ward said that it is the business of psychology 'to analyse and trace the development of individual experience as it is for the experiencing individual.' By contrast, Ribot defined psychology as an empirical, biological field – it is, above all, not 'metaphysical.' Then, in 1913, Watson, in what must be the most cited definition of the twentieth century, declared: psychology is a 'purely objective branch of natural science. Its theoretical goal is the prediction and control of behavior.' Mind, the experiencing individual, behaviour and other things besides, all have been put forward as the subject-matter.

We may perhaps tease out three questions. What is *science*? This has clearly been not just for psychologists to answer. Is science achieved by *objective methods*? A positive answer to this certainly preoccupied many psychologists. And what is the *subject-matter* of psychology as science? Answers differed.

The previous chapters gave many instances of psychologists' commitment to scientific knowledge, to science as an occupation and to the authority which, they hoped, it would command in social life. These commitments depended on belief in the objectivity and if not final truth at least rational content of what scientists discover. It is therefore all the more striking, as my list of definitions hints, to find psychologists from the late nineteenth century to the present making very different claims about the identity of psychology as science. I will not pretend to cover them all but discuss the most influential. I will include, however, an approach to science – phenomenology – which is not usually considered or, if considered, is held up by writers in English as non-scientific. Much experimental research was undertaken specifically to make psychology a science by giving it an observable and quantifiable subject-matter, knowable in the way physical objects are known. Behaviourists, physiological psychologists and modern neuroscientists went further and embraced belief that people are indeed physical objects, and hence, they thought, there is no special philosophical or religious difficulty, whatever the technical difficulties, about psychology becoming a natural science. By contrast, Wundt and Stumpf, or in the English-speaking world James and Ward, along with later gestalt psychologists and phenomenological psychologists, argued that psychology could not develop independently of arguments in the theory of knowledge (epistemology) which excluded materialism. Taking another tack, I will also look at the science of psychic phenomena (parapsychology) in order to touch on the difficulties of drawing boundaries around science.

There is an image of science as a unified body of truth. This is not in fact the case even in the physical sciences, and in psychology it is certainly not the case, although psychologists have held it up as an ideal. There have been those who have looked to a theory of function, behaviour, cognition or evolution to do the job, but also those who have looked to the unconscious, mental structure or language. We might therefore sympathize with the conclusion of Sigmund Koch's survey in *A Century of Psychology as Science* (1985): 'On an a priori basis, nothing so awesome as the total domain comprised by the functioning of all organisms (not to mention persons) could possibly be the subject-matter of a coherent discipline.'

The strength of concern about whether psychology is a science sometimes amounted to an identity crisis: psychologists desired to be objective like natural scientists but feared they were not. They studied subjective things, like feelings, or things which cannot be measured, like self-identity and motivation. Physical scientists appeared to be in the more fortunate position of studying measurable objects unambiguously external to the observer. Psychologists, who perhaps had an idealized image of the physical sciences, adopted experimental technique and mental tests in order to achieve the same stance. Later, especially in the United States, the adoption of such methods appeared a precondition for activity to count as science, even if this excluded certain topics, such as voluntary action, from investigation. At times, psychologists concentrated on the topic of learning – which in practice meant behaviour in white rats running mazes – almost to the exclusion of human thought, language, feeling or imagery. In these circumstances, psychology was distinctive as a science because of the degree to which it defined itself by its methods rather than by its subject-matter.

There always were critics, however, who denied that methods, in the absence of coherent and rational content, could ever constitute a subject as science.

We should never discount the marvellous intractability of human beings as subjects, not least because they re-invent themselves while under investigation. This, along with the psychologists' perception of the backwardness of their field in comparison with the physical sciences, explains the preoccupation with methods. Yet, by no means all research followed formal methods; indeed, much work was eclectic in the extreme. Psychology in Britain illustrates this. Right at the end of the nineteenth century, C. S. Myers, his teacher W.H.R. Rivers (1864–1922) and fellow student William McDougall joined an anthropological expedition to the Torres Straits, between New Guinea and Australia, where, among other things, they carried out psychological tests on the local people. The investigators mainly studied sensory perception and, owing to the cultural gulf between their subjects and themselves, the results had little long-term value. Myers, in the First World War, worked on acoustic perception and the detection of submarines, and after the war he moved into industrial psychology. His student, Bartlett, directed the Cambridge laboratory where he fostered research markedly different from either intelligence-testing or behaviourism. His major book, *Remembering: A Study in Experimental and Social Psychology* (1932), had no statistics and ignored behaviour. Experimental technique in Britain developed piecemeal, as a craft adapted to local circumstances and specific intellectual and practical tasks. Bartlett later wrote: 'The scientific experimenter is, in fact, by bent and practice, an opportunist . . . The experimenter must be able to use specific methods rigorously, but he need not be in the least concerned with methodology as a body of general principles.' In his view, what makes for objectivity is not a method but the ingenuity with which the investigator tests hypotheses in relation to clear questions. His were confident words, possible perhaps for the man who for many years was the only psychologist among fellow scientists in the Royal Society of London. His approach preserved conceptual analysis as part of the critical apparatus of science, and it also ruled out grand schemes to make psychology a unified subject. It exemplified the style of a small but clever British academic elite. By contrast, the Americans developed the bureaucratic, rule-bound style of a large more egalitarian profession. In England, what counted as method was a craft which a few researchers learned; in the U.S., explicit procedures put everyone on an equal footing.

## BEHAVIOURISM

Using an objective method to constitute psychology as science was the hallmark of behaviourism. Its claim was that psychology does not differ from any other natural science: it disciplines observation and builds up explanatory generalizations, and it can do this because it describes and explains the 'external' physical *behaviour* of animals or people and does not try to report on some supposed 'internal' thing like mind or a conscious state.

When U.S. psychologists in the 1960s revived discussion of cognition as a mental process, they reported a sense of liberation at feeling free to talk about conscious processes without disapproval. It suited their polemical purposes to picture psychology from about 1910 to 1960 as a behaviourist monolith. This was a false picture, not least because, before 1945, most European psychologists had dismissed behaviourism as an aberration, and because, even in North America, there were centres of research and teaching with other commitments. In addition, there was no unified behaviourist school, and in particular there were differences between the early formulations (*c.*1910–30) of John Broadus Watson (1878–1958) and the neo-behaviourism associated with the positivist theory of scientific knowledge (*c.*1930–55).

All the same, the convergence of interests on behaviour is striking. Urban and industrial society, which I have described as a condition for the expansion of psychology generally, placed a premium on knowledge of the capacities which people use to adjust to modern life. Practice and theory, influenced by the evolutionary world view, came together in an emphasis on mental *functions*, on what mind enables people to do. This 'doing' was behaviour. Measurement and testing were also concerned with performance, itself a kind of behaviour, rather than with the subjective world. In addition, psychologists felt under pressure both to demonstrate their objectivity and to claim for themselves a subject-matter separate from physiology and medicine. Colonizing the field of behaviour addressed all these needs. It also seemed to leave behind two clear weaknesses of psychology defined as the science of mind: it brought animal psychology out of a no-man's-land and into continuity with human psychology; and it side-stepped the embarrassing fact that no two psychologists could agree on the basic description of mental content – the sensations, cognitions, feels, desires or whatever, which were the supposed elements of mind. Even more, a 'can-do' attitude was, as it still is, an important strand in U.S. cultural and political life. Functionalist thought provided this with a ready-made vocabulary and underwrote it with the authority of belief in evolution. It is noteworthy that a number of important behaviourists, especially Watson, Hull and Skinner, were proud of their ability to make things with their own hands. They could fix things, and this, basically, was their model of both psychological activity and what psychologists, as specialists, do in society.

Work with animals had advantages: they were less complex, it was easier to discipline them as experimental subjects and it was possible to do things which were not ethical with humans. Watson's doctoral research, for example, involved damage to the sense organs of rats in order to study their perception. As it was not possible to ask an animal to report sensory change, as German experimenters required of people, reaction or behaviour became the data. The same point applied in studies of young children's development.

Watson's career, like many of his generation, took him from the country to the city. All the pressures to focus on behaviour came together in his world. When, in 1913, he published a concise and sharply worded manifesto, 'Psychology as the Behaviorist Views It', he gave a name and the appearance of well-defined purpose to the new direction. His views were controversial, but during the 1920s the body of opinion among American psychologists moved in his direction, albeit with refinements and qualifications, just when Watson himself moved out of academic life.

Watson grew up in a small town in South Carolina, from where he escaped to the heady modern world of the Chicago philosophy department headed by Dewey. It was difficult to adjust, and as a graduate student he found emotional support in the practical need to look after rats and build experimental apparatus for them. He completed a thesis on habit formation and quickly obtained a prestigious position at the Johns Hopkins University. Even so, he felt slighted by psychologists who treated animal work as of second-order importance, valuable to the degree that it revealed something of the evolutionary origins of the human mind but not worthwhile in itself. When he replaced Baldwin as head of department, he was able to consolidate his own programme and to gain the confidence to issue his call to arms: 'behavior men' will succeed in a psychological field which 'has failed signally . . . during the fifty-odd years of its existence as an experimental discipline to make its place in the world as an undisputed natural science'.

The negative side of the polemic was well grounded. Research on the structural elements of the adult mind had proliferated descriptions and had lost direction. In the U.S., only under Titchener's leadership was there a well-defined programme. But the constructive side of the polemic focused on what psychology could do for society, whereas Titchener thought this a task for the future. Watson did not distinguish his criticism of psychology's failure as science from his aspiration for psychology to be a human technology. Famously, he defined psychology's goal as 'the prediction and control of behavior' – the aim was visible individual achievement not an inner life or a new political order. He argued that psychologists should observe physical stimuli and responses, correlate them and thus understand, and be in a position to control, behaviour. No reference to mind, he said, should muddle the objectivity of observation. Later, he went further, and in *Psychology from the Standpoint of a Behaviorist* (1919) he effectively dismissed belief in mind as a medieval superstition. He took the tone of a no-nonsense practical man of science rather than of a philosopher, but he made the philosophical judgment that reality does not include minds. It was perhaps plausible to reject the search for a criterion of animal consciousness and speculation about animal feelings (though few pet owners agreed). But to deny mind to humans required addressing topics such as mental images, language and meaning. Watson started down this road, treating language as verbal behaviour or speech and thought as sub-vocal speech. Physiologists of an earlier generation, like Sechenov and Ferrier, who had been interested in the physical basis of mind, had made similar suggestions. In 1916, Watson attempted to record movements of the larynx during thought; but, as he was not successful, he had to be satisfied with the argument in principle. He cited as evidence the way a young child whispers while apparently thinking, and he suggested that as the child gets older this muscular movement becomes invisible and no sound is made. If psychologists were not entirely persuaded by all this, they nevertheless were attracted by the turn to explanation in terms of physical variables.

As for the philosophical problem of meaning, Watson simply dismissed it: 'We watch what the animal or human being is doing. He "means" what he does. It serves no scientific or practical purpose to interrupt and ask him while he is in action what he is meaning. His action shows his meaning.' Such statements had affinities with an influential strand of philosophy at the time, as Bertrand Russell acknowledged in *The Analysis of Mind* (1921), in which Russell developed a theory of knowledge strictly based on sense data. Russell stimulated the philosophers who became the Vienna Circle and drew their attention to Watson's work. Later philosophers, notably Ryle (though no behaviourist), developed Watson's point into an argument against the attribution of mental states to people as the explanation for what they are doing; for Ryle, mental states are dispositional. In the 1960s, however, all this came up for critical re-assessment, and non-positivist and anti-behaviourist theories of meaning and of the explanation of human actions acquired authority in philosophy and the social sciences.

There was no revolution in 1913. A change, bringing in reference to intelligence instead of reason, motivation instead of will and behaviour instead of mind had been under way for some time, and reservations and alternatives continued to exist long after. Thorndike, like Watson, had worked with animals and expressed impatience with psychology lacking social utility. Already in 1898, he had reported one of the most cited of all learning experiments, in which he observed cats trying to escape from a puzzle box. Perhaps the box was a crate and the experiments inconclusive, but Thorndike established a model for research with animal learning as the core topic of scientific psychology, and he put forward the same research as the basis for a science of education and for the control of what people do. Thorndike (and also Pavlov), more than Watson, showed how to use the precise correlation of changes in an animal's behaviour with changes in physical stimuli as the basis for a theory of learning. Thorndike summarized his own results as 'the law of effect': the consequences of movements determine whether the movements become part of an animal's repertoire, i.e. a habit. He concluded: 'pleasure stamps in; pain stamps out'. This re-described in experimental terms the utilitarian pleasure-pain principle, Locke's, Bentham's and the Mills's belief that pleasures and pains explain conduct.

Watson ambitiously advanced on many fronts at once, and the results were anything but systematic. In Baltimore he had access to the Phipps Clinic headed by Adolf Meyer (1866–1950), an influential figure in the psychiatric profession. Meyer welcomed psychology as a source of information about the conditions giving rise to illness, and Watson began to try out some of his ideas with children in the clinic. This included work with 'Little Albert', the boy whose behaviour, when he was between nine and thirteen months old, Watson and his research student, Rosalie Rayner, altered. They showed that 'Albert' (a pseudonym) had no fear of a rat; they then clanged a metal bar when the rat appeared and frightened the child. The child, they claimed, became conditioned to show fear when the rat alone appeared. If the results, according to modern analysis, were uninterpretable, the report nevertheless became a celebrated instance of the acquired nature of human behaviour. In Watson's account, it was a stimulus (the metal bar) not social structure (researches using an involuntary subject) or custom (playing with a child) which formed a person. In the 1920s, Watson and Rayner, by then married, published *Psychological Care of Infant and Child* (1928) for the popular market, spreading the idea that expertise is needed in everyday life. This is what I have called the 'formation of psychological society'. There was an audience. The *New York Times* wrote, with what appears fantastic exaggeration: Watson's *Behaviorism* (1924) 'marks an epoch in the intellectual history of man'.

There is a suggestive personal dimension. Watson's psychology sought to explain human life in terms of physical stimulus-response (S-R) links. Explaining why humans do what they do, he replaced reference to thought by reference to stimuli, and he replaced reference to will by reference to pleasures and pains, also understood as stimuli. We act, he believed, because of the pleasurable or painful physical nature of stimuli. These beliefs may have had consequences in Watson's personal life, as his medical colleague, Meyer, argued at the time. After some years in an unhappy marriage, Watson had a sexual relationship with Rayner, and during a much-publicized divorce Watson was forced to resign his university position. In letters between Watson and Meyer, Watson wrote that an unsatisfactory marriage made it natural for him to express his feelings outside marriage; the stimulus of these feelings caused his action. Meyer, however, felt that neither Watson's personal statement nor his public theory allowed for the language of mental meaning and hence for the possibility of ethical judgment. Divorce and remarriage, from Watson's point of view, were events in a pattern of stimulus-response relations; by contrast, the conservative people who disapproved of what he did looked to psychology to recognize motivation deriving from mental reflection and hence morality. Put simply: Watson did what he did and called it 'natural'; his critics thought his psychology made him blind to moral meaning.

Whatever was the case, Watson was unemployable in an academic institution. In 1921, he moved into advertising. He tried to sustain academic arguments, but his sympathies were with the practical world of quick results not with academic pedantry. After Rayner's early death in 1935, he became isolated, more at home with the animals on the small farm he built for himself in Connecticut than in the social world. Behaviourism, meanwhile, developed in far more formal directions.

Scientists and philosophers undertook to demarcate science and non-science most fiercely in the years between about 1920 and 1960. They did not succeed in producing a coherent, satisfactory general theory, however strongly they enforced boundaries in particular settings. Broadly speaking, two kinds of criteria for identifying science were explored: first, correspondence between scientific knowledge and reality, or, more accurately, correspondence between scientific statements and observation statements; second, the formal articulation of correspondence statements into unified, self-consistent theory with clearly and rigorously defined concepts. Both criteria appeared necessary. Following such general lines of argument, psychologists denied that introspection could be a source of scientific knowledge: introspective reports could not satisfy correspondence criteria, and the babble of psychological voices showed that, in fact, they had not satisfied them. The properly scientific task, therefore, was empirically to reveal the laws of behaviour, to clarify behaviourist concepts and formally to articulate consistent theory. U.S. psychologists perhaps arrived at these views piecemeal rather than through reading positivist philosophy, and certainly some, Watson himself and later Skinner, rejected theoretical elaboration in general. At times, however, there was a marked commitment to formal theory construction and the ideal of a unified science. Interestingly, there was also what amounted to obsession with unified theory in the Soviet Union. There, however, the demand clearly stemmed from each science, psychology included, having to demonstrate the consistency of its work with formally unified Marxist–Leninist philosophy.

In the 1930s, a new generation, including Edward C. Tolman (1886–1959), Clark L. Hull (1884–1952) and B. F. Skinner (1904–1990), with strict views about what makes psychology scientific, moved into academic positions. For me, it requires an act of imagination to enter the world of neo-behaviourism in the three decades from 1930 to 1960. The study of learning on the white rat became the dominant and sometimes sole topic of research, while experimenters enforced formal methods to record observations and analyse them statistically to eliminate error. (The white rat, because this animal breeds quickly, is easy to care for, can be bought off the shelf as a pure-bred strain and is a standardized subject.)

All this was part of a social process, a search for a unified field and the construction of authority for science. Even so, there was no agreement in academic psychology, let alone in psychology understood more broadly, about either basic concepts or theory construction. Both Karl Bühler and Vygotsky wrote on 'the crisis' in psychology. In the United States, where the scale of psychological activity helped make disunity visible, there were several responses. Boring published his *History*, Carl Murchison (1887–1961), who headed the psychology department at Clark University, in 1925 and again in 1930 assembled collections of position statements which juxtaposed a dozen different 'schools' of theoretical psychology. Woodworth published *Contemporary Schools of Psychology* (1931) and Edna Heidbreder (1890–1985), who had a career at Wellesley College, *Seven Psychologies* (1933). The teaching of courses on systems and theories gained an enduring place in the curriculum, in itself a way to manage the contrast between the ideal of unity and the social reality of diversity.

Diversity persisted even among learning theories. Watson's early hope that it might be possible to describe learning in terms of a small set of elementary S-R relations was quickly dashed. As early as 1922, Tolman presented a philosophically informed critique of Watson's work, in which he argued on two main fronts. First, descriptions of isolated stimuli and responses as the beginning and end of pieces of learning are artificially atomistic. Rather, Tolman argued, an animal is in an organic relationship with its environment, and a more complex, 'molar', theory of behaviour is therefore needed. Second, animals show persistence in reaching goals, and learning theory has a place for a concept of purpose. This was not intended as a return to an older psychology, and for 40 years as a professor at Berkeley, Tolman worked with rats. Debate was about how to conceptualize ordinary language's description of the rat's purposiveness in being persistent. Tolman published *Purposive Behavior in Animals and Men* (1932), which focused discussion by introducing the concept of the intervening variable in order to characterize the determinants of behaviour which lie between the independent variable (the stimulus) and the dependent variable (the response). The intervening variables roughly corresponded to what ordinary language referred to as 'mind', and research expanded. However, in the long term, it was the impossibility of extending behaviourist theory to encompass human intentionality, language and social rules which caused it to collapse.

Hull's psychology was even more formal. His magnum opus, *The Principles of Behavior: An Introduction to Behavior Theory* (1943), is mind-numbing, and it is not trivial to ask how Hull could have become perhaps the best-known psychologist of his generation in the U.S. Hull grew up in small-town Michigan and saw life as a struggle – against religion, against illness (as a student he contracted polio and thereafter walked with a leg brace which he himself designed) and against his feeling of having come to his life's work only in middle age. Both Tolman and Hull originally planned to be engineers, and Hull retained in his psychology an engineer's imagination about mechanical relationships and an engineer's belief in the possibility of constructing complete working systems. Hull even invented and made a machine to calculate correlation coefficients in the mid-1920s – a practical demonstration of machines being able to do what mental processes do. He never ceased to think mechanistically about both human beings and organizing productive research. Hull sought certainty and rigour, which were qualities he found in geometry and in Newton's great work: every visitor who went into Hull's office in the late 1930s encountered an open copy of the *Principia*! In Hull's view, psychological science is a formal hypothetico-deductive activity. When the scope of Pavlov's work became known in English in 1927, Hull perceived conditioning experiments as the way to study all aspects of thought in terms of habits. Psychology, he proposed, will become a science when it uses facts about habits to postulate provisional axioms, deduces consequences from these axioms, tests the consequences in experimental settings and finally uses the results to refine the axioms into a conclusive form.

All this might have remained fantasy. In 1929, however, Hull moved from his job as a teacher at the University of Wisconsin to a research position at the newly founded Yale Institute of Human Relations. This institute, funded by Rockefeller, had the mission to unify social science research. Adopting the model established in medicine, the plan was to bring together people from different disciplines to work in groups on integrative tasks. Dissatisfied with lack of progress and with researchers who continued to work individualistically, the Rockefeller Foundation threatened withdrawal. In this situation, Hull seized the opportunity to head a co-ordinated programme in which he shaped the axioms and general theory while groups of researchers, organized in a hierarchy, worked out the detail through experimental studies. He was a man in a hurry, anxious to axiomatize serial habit formation as the basic law of social science, and he generated a mass of empirical questions and points of detail to be answered later. Following the Depression, academic funds were short, and Hull's programme stood out in the 1930s as well organized, purposeful and well funded. Hull appeared the man to bring psychological theory up to the standard of the physical sciences.

In pursuit of deductive rigour and unity, Hull's system never achieved either. It simply collapsed because of mismatch between aspiration and execution and the weight of its own complexity. His work survived his death in 1952 only in the less deductive and more restricted version propagated at Iowa by Kenneth Spence (1907–1967). In addition to the problems of creating a formal theory, there began to be vocal critics, led by Koch, of the belief that learning is a unified activity dependent on reinforcement. Learning, critics claimed, is not a single process but encompasses many things. The clearest and most embarrassing evidence for this was the way psychologists' views about what learning is varied with where they had trained. The joke that it was perhaps experimenters rather than rats who learned a particular performance was not lost on observers.

One prominent position, linked to neo-behaviourism by methods and ideals, remained uninfluenced by criticism, and indeed, during the

1950s and later, flourished as a distinct school. This was Skinner's operant psychology, which had a firm institutional base first at Harvard and then elsewhere. Skinner's psychology was not simply another form of neo-behaviourism, since he rejected completely the value of theory and formalized knowledge. He rejected theory construction for induction, generalizing only from observed instances. His psychology was a re-run, though with more rigorously defined experimental procedures and a different learning theory, of the anti-mentalist zeal of Watson.

In his first book, *The Behavior of Organisms* (1938), Skinner attacked the notion of trial-and-error learning and instead argued that all behaviour involves the one basic principle of reinforcement. An animal, he argued, makes movements, some of which are reinforced while others are not. He demonstrated this using his favourite set-up: releasing grain to a pigeon when it raises its head above a certain line. A pigeon which raises its head, if followed by the release of grain, as a matter of fact then raises its head more often. If this is so, it should be possible to devise schedules of reinforcement, that is, a programme to induce one pattern of behaviour rather than another. His descriptions in principle did not refer to mind (e.g. pleasure and pain), to drives (e.g. hunger), to habits (e.g. learned ability) or even to stimuli (e.g. the appearance of grain). He recorded sequences of movements and argued that science consists solely in the description of these sequences. This was indeed radical.

Skinner consistently maintained a number of principles, expressed in *Science and Human Behavior* (1953), and this very consistency made him a marked figure. Since he denied the value of theory and argued that all science needs is systematic observation, he usually declined to debate his position and instead presented his empirical observations as his position. Second, he separated psychology from physiology and thought the latter a source of misdirected experimentation and speculation in psychology. While he accepted, of course, the presence of internal physiological factors initiating animal movement, he thought this of no concern to psychology, the business of which is to observe movement. He also refused to refer to mental states and to ethical concepts, like freedom, usually associated with mental states. Animals and people alike exhibit a pattern of movements with operant value, nothing more. Thus, for example, 'we may regard a dream, not as a display of things seen by the dreamer, but simply as the behavior of seeing'. Finally, he took biological adaptation to be self-evidently the basic value of human life. He turned this evolutionary value into a fully fledged theory of culture, though, in Skinner's terms, he neither held a value nor advanced a theory but simply made inductions from scientific observation.

Science shows how we can survive, and only a dinosaur, it appeared to him, would not embrace operant science.

Skinner's ability to stir up controversy because he worked out his scientific position to its logical conclusion came to a climax with his best seller, *Beyond Freedom and Dignity* (1971). He attacked notions of free will and individual responsibility as remnants of a pre-scientific past which were leading to social and political disaster. This was not the embittered talk of an old man but the utopian dreaming of one young at heart. In his novel, *Walden Two* (1948), Skinner had imagined a community run on radical behaviourist lines and achieving harmony and efficiency impossible in the outside world. A few such communities modelled on Skinner's ideas even existed. He had also constructed a baby box, his model of a good environment, which earned him some notoriety because it appeared so unaffectionate, though the box was little more than an easily cleanable crib with controlled conditions, in which his daughter, Deborah, lay naked and in comfort.

Skinner never denied the existence of consciousness, though he thought it nothing but the accompaniment of certain physiological states of no concern to psychology. This, as he perceived, made the ability of his psychology to explain language the test case of everything he had worked for. *Verbal Behavior* (1957) was the outcome of a twenty-year effort to deal with this, and the book argued that language consists of reinforced movements like any form of behaviour. When we say 'red', for example, we have a physiological state (at present unknown) and we make movements of the larynx which, earlier in life, was reinforced by other people's use of the same movements. If a child says, 'red', and parents smile and say, 'yes, red', the child says it again in the same circumstances. Noam Chomsky reviewed the book, and his scathing attack firmly set linguistics on another path and convinced many that Skinner did not have a unified science. By the 1960s, psychologists trained in Skinner's experimental methods were a large and productive group but somewhat isolated. His techniques left a mark in machine-learning, as used in language laboratories, and in training people, for example, using token rewards to reinforce habits of cleanliness in children.

Criticisms of neo-behaviourism in general became common in many areas of psychology, including perception and child development, as well as language theory, in the second half of the 1950s. A vision of all learning conforming to a single pattern, and even more a vision of learning as the key to psychology's secrets, appeared chimerical.

## PAVLOV AND 'HIGHER NERVOUS ACTIVITY'

Part of the project for psychologists studying behaviour was to mark out a field separate from nervous physiology. Questions about psychology's relations to physiology did not go away but, all the same, at the beginning of the century there was something of a move to distance psychology from physiology, since the brain was exceedingly difficult to study, little was known and progress might come more readily from psychological research. This was Watson's view, for instance. There were certainly psychologists, like Piéron, who promoted physiological psychology as a speciality, but for most psychology's credentials as science depended on other areas of work. As for physiologists themselves, they were generally too busy to be distracted by the enigma of mind's relation to body. Only after 1945 did things change.

The scientific programme of Ivan Petrovich Pavlov (1849–1936) is an interesting and influential exception to these generalizations. He consistently denied being a psychologist rather than a physiologist; yet his research claimed the kingdom of scientific psychology. This needs explaining.

Pavlov was the son of a priest who, like many ambitious and idealistic students of his generation and background in Russia, turned to the natural sciences. The virtues of science shaped his life. He studied medicine in St Petersburg and then continued with physiological research at the university in Breslau, leading to the work on digestion which brought him the Nobel Prize in 1904 and world renown. In middle age, as a professor at the military medical academy in Petersburg, Pavlov switched his full-time attention to psychological events and to the brain as their material basis. He continued to call himself a physiologist because, for him, *scientific* explanation requires knowledge of material, physiological causes, whatever the psychological subject-matter. He contrasted objective physiological science and subjective psychological speculation, and this gave direction to all his work.

The research itself built on the basic procedure of training dogs to salivate at the sound of an electric buzzer (not a bell). Normally dogs salivate at the smell of food, and the research studied, under varied conditions, the way dogs salivate at a sound when food has previously followed the sound, that is, when they have 'learned' something about what precedes food. In a rhetorical reconstruction of the key step in developing the research, Pavlov wrote: 'I decided . . . to remain in the role of a pure physiologist, *i.e.*, of an objective external observer and experimenter, having to do exclusively with external phenomena and their relations.' He therefore did not describe the dog's experience but referred instead to the unconditional reflex for salivation, the innate response to food, and to the conditional reflex, the trained response after a sound has become a stimulus for the reflex. On the back of this, Pavlov designed a research programme on the psychological topic of learning in terms of the physiological events of conditioning. He viewed conditioning in physiological terms and claimed to connect conditioning processes with nervous events in the brain. This left no space for an independent science of psychology.

Over 30 years, till his death in 1936 as a formidably energetic and autocratic old man, Pavlov directed an ever larger group of scientists who developed this research. In his earlier research on digestion, Pavlov had mastered the organizational skills necessary for running a large laboratory and bringing together many students (largely doctors in training) to work on a co-ordinated problem. He launched a large-scale research programme, which soon differentiated conditioning and deconditioning into many sub-varieties, with studies of inhibition as well as excitation. Pavlov inspired younger researchers, including a significant number of women, with a belief in conditioning as the objective method to study events previously attributed to mind. It became possible, for example, to analyse a dog 'seeing' by conditioning and deconditioning dogs to different stimuli. The researchers explored differential responses rather than making claims about what animals see, replacing statements about the dog's subjective world by objective observations. In one of the best-known researches, N. R. Shenger-Krestovnikova induced artificial neurosis in a dog, which she trained to salivate in response to a circle and not an ellipse and then presented with intermediate shapes. She and Pavlov interpreted this as a demonstration of the conflict between excitatory and inhibitory processes in the brain and not as a mental event. In such work, Pavlov pushed forward his ambition to bring human beings fully within the scope of natural science. He extended his arguments, though in fact speculatively, to human mental disorders, character types and individual differences, language and the emotions.

Pavlov himself always intended to go beyond description to explain at the level of physiological processes in the brain. He therefore called his science 'the theory of higher nervous activity' and planned for it to occupy the place filled elsewhere by psychology. This larger project, however, did not make progress on the world stage. For one thing, though Western psychologists knew Pavlov's work on conditioning from 1906, they regarded it as a valuable experimental technique; they did not know about or understand its status for Pavlov as the route to a new science. For another thing, when his lectures on conditional reflexes became available in English translation (in two translations in 1927 and 1928), Pavlov's elaborate correlation of the varieties of conditioning, deconditioning, inhibition and so forth with higher brain activity appeared crude to Western physiologists. In a situation where English-speaking neurophysiologists, following Sherrington, studied neuronal activity, Pavlov speculated about the large-scale irradiation or spread of excitation and inhibition across the cortex (the whole structure of the higher brain). The division of labour in Western science, with few exceptions, had fostered highly specialized research and created barriers between the disciplines of neurophysiology and psychology. Held up beside this, the Pavlovian bid to unify a science of human nature in terms of brain activity appeared crude and, indeed, without interest. As a comprehensive alternative, Pavlov's programme remained restricted to the USSR, at least until the Russians imposed it throughout the Soviet empire around 1950.

From what they knew of his methods, behaviourists claimed Pavlov as one of their own. Pavlov, however, vehemently and correctly rejected this, and he even made the long journey across the Atlantic in 1929 to say so. It was, for Pavlov, not the method of conditioning but the substance of knowledge of higher brain activity which constituted science. American psychologists, however, subsumed his research into the debate between the two main types of learning theory. On the one side, there were those who followed the utilitarian philosophers and Thorndike and understood learning as the consequence of reinforcement. On the other side, there were those, like Pavlov, who attributed learning to the contingent relations of stimuli in the organism's environment. Competition between these two basic models, not anything resembling a theory of brain activity, set the agenda for half a century in the U.S. In the 1930s, Skinner elaborated a purified form of the reinforcement theory, while Edwin R. Guthrie (1886–1959) further developed the contingency theory.

In 1916, in a rare comment on his general stance, Pavlov dismissed, from science, a concern with mind and consciousness: 'We are studying all the organism's reactions to events of the external world. What more do you need? If you find it nice to study, so to speak, the poetry of the problem, then that is already your business.' Nonetheless, Pavlov was highly cultured and no materialist. What he did possess, always in public if not perhaps always in his quietest moments, was a faith in science, and it was this faith which gave him the energy to maintain a research community through the years of war, civil war and extreme material hardship in Russia from 1914 to 1923. As a world-famous scientist, Pavlov had long been in a position to campaign for funds to extend his work, though he had trodden carefully under the tsars when there was pressure not to question the divine essence in humans or weaken belief in individual responsibility. His faith in science and world position also stood him in good stead when there was fierce competition for scarce resources during the 1920s. The Bolshevik government supported Pavlov's work financially, initially because it wanted Pavlov to stay in Russia and to have the prestige associated with the work of a scientist of international stature. Later there was also support because some theoreticians thought it possible to harmonize dialectical materialism and Pavlovian theory; more importantly, Pavlov could be pictured as a home-grown hero.

Until the mid-1920s, though, both the Moscow psychological institute and Bekhterev's psycho-neurological institute in Leningrad (formerly St Petersburg) competed with Pavlov for state support and leadership in the objective science of man. Bekhterev's programme, under the general heading of 'objective psychology' and later 'reflexology', finally failed to attract political backing. Bekhterev rather vaguely used the word 'reflex' as an inclusive term in a theory explaining the organic connections between a person and surrounding conditions. His programme was, like Pavlov's, biologically oriented, and neither programme accorded psychology the status of a subject independent of physiology, albeit in the West their reputation was as promoters of objective psychology. Bekhterev's programme differed by including more neurophysiology and clinical neurology, and more specifically by rejecting Pavlov's method of conditioning as a way to uncover knowledge of the brain. By the late 1920s, however, Pavlov's programme flourished alone, at the expense of other research in Russia, and it had a level of support which areas of psychology independent of his work did not.

At first glance it is surprising that Pavlov emerged as a leading scientist in the Soviet state. He initially had nothing but contempt for what he called the 'barbarians' who seized power, but he was so committed to natural science that he was prepared to judge the new regime simply by whether or not it supported his work. As explained, the Communist Party had its own reasons to back a scientist of Pavlov's stature. Moreover, in 1924, Nikolai Bukharin, then the Party's chief ideologist, broadly endorsed Pavlov's work as 'a weapon from the iron arsenal of materialism' and, as long as the meaning of the slogan was not gone into too closely, this added support. Consequently, by the early 1930s, when Pavlov was in his eighties, he was head of two purpose-built institutes, one in Leningrad and one, the Koltushi biological station, in the nearby countryside, which together employed some 40 scientific workers. His comments about communists mellowed, and in 1934 he offered public support to the state, though the content of his science never included any reference to Marxist principles. By the time of his death in 1936, Pavlov was described publicly as a hero of socialist labour.

## THE SCIENCE OF 'THE PHENOMENON'

During the so-called age of behaviourism, Spearman in London, Bartlett in Cambridge, Pavlov in Leningrad, Piaget in Geneva and the Bühlers in Vienna carried out very different research. This was also the case in Berlin in the interwar years, where gestalt psychology ruled. This was not just another school of psychology, though activity there did have an unusual degree of institutional and intellectual coherence. It was a full-bloodied attempt to establish psychology as a science by re-examining the foundations of the sciences in general. The three architects of gestalt psychology, Max Wertheimer (1880–1943), Kurt Koffka (1886–1941) and Wolfgang Köhler (1887–1967), shared the experience of cultural crisis which gripped European intellectuals before and after the First World War. They brought to psychology a debate about science and values thought crucial to civilization itself. It is not that North Americans (like Cattell or Thorndike) did not also think in grandiose terms, but, for them, this expressed itself in seeking to apply psychology to ensure social progress. The Europeans were concerned about philosophical fundamentals, and their style of work was in marked contrast to much which went on in the U.S. This contrast, in personal manner as well as style of thought, became a visible divide after the gestalt psychologists migrated to America.

The European background was a concern with securing the basis for philosophical knowledge of what it is to be human, and hence for knowledge of a being with values as well as existence, for the life of the good, the true and the beautiful – the life, as many wrote, of 'the human spirit' – as well as biological life. Around 1900, many thought philosophy had failed in this task, a failure highlighted by the success of the natural sciences and human sciences (like history and philology) in establishing knowledge in empirical terms but not in terms of comprehensive reasoning. Radical innovations in physics – Einstein's theories of special and general relativity (1905 and 1916) and the quantum mechanics of the 1920s – questioned the intelligibility of causal understanding itself. Looking beyond the university, scholars feared modernity, and industrialization had brought about a shift of power to business and to the technical professions away from a cultural elite. Various new directions were taken, but the relevant one for psychology was a return 'to the phenomenon', that is, to a new and accurate description of the conscious world existing logically and empirically prior to knowledge of other things. This kind of argument appeared, in very different ways, in Brentano's psychology 'from an empirical standpoint' and in Mach's positivism. Another version, with considerable twentieth-century influence following the work of Husserl, became phenomenology.

Conventional histories of psychology tell how the field became scientific, which, in English-language usage, refers to the way psychology followed the example of the natural sciences. This excludes phenomenological psychology from the story. An alternative story, however, might concern itself with psychology becoming a science by laying a foundation 'in the phenomena', the conscious world, not in the kinds of objects the natural sciences presuppose, like brains or chemicals, or indeed not in the kinds of objects idealists presuppose, like the self or the Absolute. A scientific psychology, with this viewpoint, existed, aiming to describe the phenomena of the conscious world objectively, without presuppositions. Phenomenology underwrote hopes for scientific psychologies quite unlike psychologies modelled on the natural sciences, such as behaviourism and the theory of 'higher nervous activity'. Psychology, phenomenologists argued, cannot take the methods and entities of natural science as its starting point but must start from the lived knowledge we have in conscious awareness, in all its qualitative and evaluative multiplicity.

Husserl's *Logische Untersuchungen* (1900–01; *Logical Investigations*) set out to make a new start. The book's dedication to Stumpf, however, pointed out the manner in which this new start rested on work already undertaken to describe the qualitative character of conscious awareness. Dilthey, for his part, having read Husserl, returned with renewed vigour to describe the psychic nexus, the structure of living and inherently purposeful conscious life – a conception of which existed in his earlier approach to psychology. Husserl called his project 'phenomenology' as he used logical procedures to examine the conscious or phenomenal world in terms proper to this world. His project was not to be more objective in introspection but escape from the subject–object distinction in order to describe what 'is': 'Being'. He inspired others, including Heidegger and through him Sartre, to turn language describing phenomenal consciousness into language characterizing existence. Down this road, his work developed with little contact with the natural sciences. Nevertheless, a number of psychologists, supported by an educated public, hoped to find in phenomenology grounds for a psychology which would do justice to the spiritual, purposive and value-asserting side of human life, the side which natural science appeared to ignore or even rule out. Husserl, for example, wrote: 'Experienced sensation is besouled by a certain act character, a certain grasping or mean-ing.' Such expressions – 'besouled', 'grasping or mean-ing' – resonated with readers who feared the natural sciences excluded from knowledge the reality of what matters in life.

Husserl's interest in the phenomenal world derived in part from Brentano's lectures. Husserl rejected Brentano's characterization of the mental and instead, more radically, sought to 'bracket-off' the conventional attribution of conscious awareness to mind in order to describe conscious awareness 'in itself', and he sharply criticized the Cartesian dualism of mind and body. Phenomenologists sought to shut out the contingent attribution of conscious qualities to things (for example, attributing colour to a physical thing) in order to describe their nature and value (the colour as a state of being). Brentano had also influenced Stumpf, who, though he did not teach in philosophical terms, provided institutional support for a psychology focused on phenomenal qualities. Among English-language writers, James, with his uniquely colourful language, gave some idea of the gulf between the riches of the directly experienced world and the desiccated data in terms of which experimentalists turned that experience into the subject-matter of natural science. In France, as James was aware, there was a language going back to Maine de Biran which sought to go beyond perception of passive nature to do justice to the living, desiring and immediately perceived world of the self. Literature, poetry and the arts generally, it scarcely needs saying, had long provided a home for the phenomenological imagination.

If Husserl's writing was philosophical, others turned to construct a phenomenological psychology, and this was connected to experimental projects both in the Würzburg school and in the work of the gestalt psychologists. In Würzburg, in the decade before 1914, Karl Bühler's analysis of complex thought and the work of Narziss Ach (1871–1946) on volition rejected the correlation of elementary conscious states and simple external variables. The experimenters instead described the conscious world in order to represent its intrinsically active structure. It is worth remembering that all German-language psychologists had been students of philosophy, and they took for granted a conceptual language which left many psychologists in the English-speaking world, if they attended at all, simply flummoxed. Sometimes Husserl's influence was direct. David Katz (1884–1953) studied at the university of Göttingen, where Husserl taught, and he went on to teach at Rostock and then in Stockholm. He worked at length and with originality on the richness in experience of colour and touch, finding ways to describe this richness as part of an experimental programme. The titles of his main publications reflect what he himself called a 'phenomenological method': *Die Erscheinungsweisen der Farben und ihre Beeinflussung durch die individuelle Erfahrung* (1911; literally, 'the modes of appearances of the colours and their modification by individual experience', published as *The World of Colour*), and *Der Aufbau der Tastwelt* (1925; *The Structure of the World of Touch*). Describing colour, for example, Katz distinguished film colour and surface colour; the latter, in conscious experience, attaches to a surface, the former not. In his account, these two experiences are not modes of appearance of one and the same colour but are different colours; which colour we see depends on context. His language drew qualitative distinctions between lustre, glow, luminosity and the like, familiar to painters but previously outside the range of objective research. In his book on touch, Katz pointed to the 'almost inexhaustible richness of the touchable world'.

A key writer in the creation of a wider audience for phenomenology was the Munich philosopher, Max Scheler (1874–1928). During the 1920s, he linked description of the structure of consciousness to public debate about values, and his books appeared to reclaim science from materialism and fulfil a human need for meaning. Such work subsumed psychology within philosophical anthropology and discussed human nature not as a matter of biology (though some authors incorporated the biology of human nature) but of each person's irreducible value as a concrete realization of the general 'Being' of humanity. Scheler wrote especially on the emotions, discussing them not as subjective feelings or physiological events but as universal conscious structures objectively and meaningfully connected to the values to which the emotions refer. Pursuing this, he argued that moral action is not a matter of subjective sympathy with others, not a matter of feeling per se, but a matter of the objective orientation of consciousness, as our essential nature, in an authentic manner.

Many readers during and after the moral and material catastrophe of 1914 to 1918 looked to authors like Scheler to renew hope. Such philosophers appeared to authenticate emotions and values as the precondition for knowledge in psychology. The endemic danger was pretentious and vague language, rhetorically summoning up thought healing the human condition while in fact giving it little precise content.

There were links with medicine. Some psychiatrists took up phenomenology and described mental illness in terms of disturbance of the evaluative structure of phenomenal awareness (in which, for example, a normal feeling of the separateness of existence becomes an acute feeling of alienation). Developed systematically in *Allgemeine Psychopathologie* (1913; *General Psychopathology*) by Karl Jaspers (1883–1969), the patient's conscious orientation towards the world, rather than morbid biology, became the starting point for knowledge. Scheler himself contributed descriptions of self-deception and of *ressentiment* (a feeling of envy and resentment and hence distance from others) which influenced psychiatry in Munich. In the 1950s and 1960s, phenomenology also became an important resource for humanistic critics of behaviourism and of natural science models of human life and mental abnormality in the English-speaking world.

French, Belgian and Dutch intellectuals took up phenomenology with interesting consequences. An example is F.F.J. Buytendijk (1887–1974), who, from 1946, was a professor at a large institute for psychology in Utrecht. Buytendijk, a physiologist and a physician, was interested both in Scheler's philosophical anthropology and in the life of animals. He wanted to know human beings in what he called their 'innerness'; similarly, he wanted to observe animals as they express themselves in their distinctive ways of life, and this, he argued, requires studying natural habitats not experiments. Knowledge of animals, he thought, presupposes our own experience of the human situation. Animal studies must have a foundation in general psychology, and general psychology must address the human existential condition – the meaning being in the world has for each person.

According to Buytendijk, we gain real knowledge of other people, that is, knowledge of their 'innerness', through the 'encounter', which involves 'disinterested and desire-free yet personally interested participation in each other'. Referring to the then newly fashionable American expression, 'OK', which he deplored, he wrote: 'fashion is not a habit, but a significant behavior, the expression of a value-project on the part of human being as being-in-the-world. . . . We can only guess at the influence which the fashionable expression "O.K." is having on contemporary European society.' This expression, he feared, substituted unfeeling mechanistic relations for 'encounter'. He translated such values into a wide range of detailed descriptions of human movement (noting, for example, the special meaning of self-movement and the difference between touching and being touched); similarly for pain. Dutch psychologists influenced by Buytendijk described the meaning in a person's awareness of everyday psychological events like shaking hands and driving cars. The methods were far from achieving the kind of objectivity enforced in English-language psychology, but for phenomenologists they were objective because they rested on values intrinsic to the human condition. Indeed, the Utrecht psychologists considered their psychology to be more objective than any psychology based on natural science and the denial of inner nature could ever be. Nonetheless, by the late 1950s, the tide in the Netherlands had turned against this view.

Phenomenology appealed to intellectuals seeking to unite knowledge and values. In Spain, for instance, before the period of the Franco regime (which began in 1938), the philosopher of 'vital reason', Ortega y Gasset (1883–1955), included reference to it in his efforts to open society to wider European culture. The group of people who formed around Ortega, the so-called school of Madrid, associated itself with Husserl but characterized 'Being' in terms of life acts rather than following Husserl's radical reduction of consciousness. This elevated the individual's subjective awareness of meaningful action to the key position in an idealist, but still psychological, analysis of the human condition. In France, phenomenology entered philosophy through a variety of routes but had little direct influence on those who were primarily psychologists by training. In the late 1930s, Maurice Merleau-Ponty (1908–1961) and Jean-Paul Sartre (1905–1980), both philosophers, wrote on the emotions, imagination and other topics from a non-experimental and phenomenological, rather than psychological, point of view, claiming objectivity for their descriptions of universal features of the human condition. As Sartre's own turn to writing stories, plays and novels indicates, their approach was close to literature. Merleau-Ponty's work, however, which was more rigorous and formal, did produce a major critique of the psychology of perception, behaviour and personality in *Phénoménologie de la perception* (1945; *Phenomenology of Perception*). In a characteristic phrase, he argued for a 'turn back to the phenomena': there can be no other starting point for philosophy and psychology than 'presence' as part of the world, than the meaning which consciousness has for the embodied conscious person. His rejection of mind–body dualism and description of psychological life as intrinsically embodied was of particular interest to psychologists. There is, he argued, no observer of a physical world, body or behaviour; there is, rather, meaningful embodied being, given in the act of attention, and psychologists should therefore analyse perception as action in the world, as motor as well as sensory activity.

Phenomenological psychology became known in the U.S. and Britain through the work of psychologists such as Robert B. MacLeod (1907–1972), but, excepting humanistic psychology and psychotherapy, it had little impact. It was completely unacceptable, a form of reactionary idealism, in the Soviet Union. For most psychologists, East and West, the quest for a science of psychology was a quest for a natural science.

Gestalt psychology, in this situation, is historically of great interest. Though psychologists in this school also wanted to return to 'the phenomena', and though some, notably Köhler, had large philosophical ambitions, they firmly intended to integrate their psychology with natural science. There were a number of precedents (for example Cousin, Wundt, Mach) for developing empirical psychological resources in the service of philosophy, and the gestalt psychologists followed this pattern of argument. Their special contribution was to describe conscious structures as organized wholes or '*Gestalten*' (literally, 'forms'), understood as natural reality. They thus opposed the tradition from Locke to Mach and Titchener which represented the basic elements of mind in terms of sensations (or, in Hume's language, 'ideas') conceived as analogous to physical atoms, and which conceived of knowledge as consisting of either associations between these sensations or apperception, the mind's organization of sensations.

The well-known case of a gestalt phenomenon is the line drawing which looks like a duck at one moment and a rabbit the next. We do not see a series of lines, assemble an image of a duck and then re-assemble the lines as an image of a rabbit. The image is real on the page as a physical thing and in consciousness as a form. It was Wertheimer who established the exemplar for the gestalt approach to perception in a paper, published in 1912, on an already known instance of apparent movement. He flashed two stationary light sources at intervals; with a certain timing of the intervals, the observer will see one moving light. Though this at first glance may appear an obscure phenomenon, it both presented a great challenge to understanding and suggested a way experimentally to tackle it. Wertheimer argued (he called this the 'phi-phenomenon') that the observer perceives *motion* as a dynamic psychological event or whole. The mind (or brain), he argued, does not turn sensations into perceptions; rather, consciousness itself has an organized temporal and spatial structure. We do not perceive events in space and time, but space-time is the given structure of conscious processes. Last, though the gestalt psychologists only later stated this as a general principle, Wertheimer presumed that the psychological, mental organization of forms is strictly parallel to the brain's physical organization. The scientific programme for psychology was, therefore, through experiment, to find the mind–brain structures.

Wertheimer was born in Prague into a family which assimilated German-language culture in public while maintaining Jewish traditions and learning at home. He then rejected Jewish beliefs for humanism, while acquiring, as did Koffka and Köhler, knowledge of experimental psychology from studying with Stumpf. Like Stumpf, Wertheimer, a pianist and violinist, was intimate with music. Music – in which phrasing, rhythm and harmony exist as wholes and as aesthetic values not assembled piecemeal by the player or listener – was an aesthetic model for gestalt psychology. While behaviourists like Hull and Skinner constructed mechanical devices, Wertheimer and his friends played Haydn quartets. Koffka was attracted to Stumpf's psychology as it appeared to him to be a science relevant to daily life, and daily life, for Koffka, included his own partial colour blindness, which was one topic of his research. After Berlin, he went to Würzburg, where he learned about complex experimental studies of thought and mental acts and worked with researchers who treated higher mental processes as dynamic events rather than as entities (like intelligence). Köhler's father was director of a *Gymnasium* in Saxony. Attracted to the physical sciences in Berlin, Köhler studied sound pitch in Stumpf's institute.

These three young men, who shared a cultural background, training in experimental techniques for the precise description of conscious awareness and exposure to the debates dividing psychologists, came together in Frankfurt in 1910. The story involves a fine myth – it may even have happened! – about scientific insight. While travelling on a train through Frankfurt, Wertheimer got so excited by an idea for an experiment on apparent movement that he left the train in order to buy a toy stroboscope (a device for flashing light). He played with this in his hotel room and then descended on Köhler, a teacher in the psychological institute of the academy of social and commercial sciences (subsequently the university) in the city. Wertheimer showed with his stroboscope, Koffka later wrote, that 'movement as experience is different from the experience of successive intervening phases'. Wertheimer and Köhler were soon joined by Koffka and his wife, Mira. They liked each other and were fascinated by the same intellectual problems. They shared belief that insight into the organized form of consciousness gave them the basis for making psychology systematic and hence scientific.

The researchers did not limit themselves to perception. Even before his work on apparent movement, Wertheimer had written a paper on the numerical capacities of 'primitive' people, suggesting that they represent numbers as part of an organized way of life. Thus, he proposed, 'primitive' thought about numbers is dependent on the mutual relations and context of things rather than on a notion of arithmetic number. This explains, he argued, Western observers' reports that people 'fail' with elementary arithmetic. His conclusion was that knowledge of psychological life requires research into the situation: it cannot begin with assumptions about the natural status of calculations with elementary entities. Köhler reached a comparable conclusion with intelligence. In 1913, he went to Tenerife in the Canary Islands, intending to stay for a year as director of the recently established Prussian Academy of Sciences's animal research station. However, he had to stay until 1919 because of the war. While there, he carried out experiments with chimpanzees which made his international reputation. Köhler observed the animals learning to obtain food rather than testing for their intelligence, and he observed insight. In contrast to U.S. learning theorists, who set animals tasks like pressing a bar, which dictated either a unit positive or a unit negative response, Köhler gave his animals situations in which they could act in complex ways. He then saw chimpanzees put sticks together or pile up boxes in order to reach bananas which were otherwise out of reach; sometimes he believed he saw sudden insight before the chimp purposefully solved the problem. For Köhler, this was convincing evidence of animals learning by the comprehension of a situation rather than by assembling habits acquired through trial and error. Given the focus in the United States on learning as the topic with the potential to unify psychological research, and given that this research claimed unique scientific credentials because it linked objectively observable stimulus-response elements, Köhler's challenge to it attracted attention and controversy. The contrast and sometimes conflict between gestalt and U.S. learning theory lasted over several decades.

In 1920, Köhler began to teach at the university in Berlin and then, in 1922, he was called to Stumpf's chair on the latter's retirement. From this position, in which Köhler became very much an academic aristocrat in the German manner, and with Wertheimer's collaboration in Berlin and then in Frankfurt and Koffka's support from a chair in Giessen, Köhler directed an extremely creative and productive group of students and colleagues. The Berlin group included the brilliant young psychologist Kurt Lewin, and philosophers of the calibre of Ernst Cassirer and, later, the theorist of probability, Hans Reichenbach, associated themselves with its work. The core of the Berlin research was technical experimental studies, and gestalt-oriented researchers achieved leadership in research on perception. Köhler, who had the competence to work in mathematics and physics and take the arguments to the wider community of physical scientists, ensured that the larger philosophical project remained in view. Ordinary experience and scientific experiment were to demonstrate the inherently organized structure of mental content, the isomorphism (i.e. the same form) with brain organization and the physical realities of structural organization in systems ('fields') in general. Even more ambitiously, the researchers contemplated the analysis of moral and aesthetic values as part of the phenomenal field and hence the subject-matter of science. They never lost sight of 'the crisis' of knowledge which some had blamed on nineteenth-century mechanist science. For the gestalt psychologists, imbued with the culture of old Europe, positivism and behaviourism capitulated to irrationality.

Irrational beliefs, however, acquired political power in Germany in January 1933. Köhler was a rare example of a senior academic who publicly objected to the dismissal from the universities of people whom new laws classified as 'politically unreliable' or 'Jewish', and he wrote a newspaper article in defence of Jewish colleagues. According to his own account, he expected to be arrested and sat up all night with friends to play chamber music. The arrest did not happen, though his institute was harassed and his assistants dismissed. Köhler, who conceived of himself as a scholar above politics, continued to argue with the authorities and to help his staff, and he began his lectures, as required, with what everyone knew to be a token Hitler salute. In 1935, he finally resigned, having decided that the conditions were incompatible with scientific work, and he left for the United States.

Koffka had been attracted to Smith College, Massachusetts, in 1927, and Wertheimer, in exile, taught at the New School for Social Research in New York, the independent institution, founded in 1919 for 'educating the educated', which gave much support to refugee academics. Thus the gestalt movement reassembled in the United States. Homegrown psychologists, however, perceived it to be a school with a theoretical orientation relevant to studies of perception, not a philosophical movement, and they debated its experimental work without reference to the conceptual framework. The gestalt psychologists did not gain prestigious institutional positions or attract substantial numbers of graduate students to perpetuate their work. In competition with other 'schools', gestalt psychology lost out. The challenge which gestalt psychology posed to mechanistic thinking barely crossed the Atlantic.

## PARAPSYCHOLOGY

Behaviourists and Pavlovians alike rejected metaphysics and went to great lengths to claim knowledge with the authority of rigorously empirical observation. They proffered training in their respective experimental methods as the route to science. They polarized reason and unreason, psychology and superstition (as in the epigram from Watson about getting rid of 'medicine men'). The business of policing the boundary of scientific psychology promoted an occupation, denigrated a pre-scientific past and outlawed reprehensible manifestations of that past in current beliefs. For some people it simply boiled down to science versus religion. A more immediate target for psychologists was belief, exemplified in astrology and spiritualism, which put forward psychological claims as if with empirical authority but which, as the psychologists saw it, in fact pitched pseudo-science against science. A sketch of the history of parapsychology will illustrate this. Most psychologists, of course, never thought about parapsychology (the study of extra-sensory perception, ESP); they dismissed it to the realm of the impossible or at least the unthinkable, of interest only to cranks or, more generously, to those who like such things. Yet a small minority gave it attention and raised interesting questions about what counts as objective evidence and what separates science and pseudo-science. Research on parapsychology, unlike phenomenology, did not seek to break the mould of natural science research but hoped to bring a new range of phenomena within the scientific world view.

Modern parapsychology originated in the Victorian enthusiasm for spiritualism, which was a search for personal meaning and comfort from a spirit world and knowledge of life after death. Like other 'fringe' movements, such as Creationism, spiritualism also had the potential to give a voice, even power, to individuals and ways of life not certified by 'the establishment'. Victorian doctors and physiologists, all men in this connection, linked trance states, table-turning, spirit materialization and the like with the hypnotic sleep produced by healers and entertainers. Medical psychologists studied these phenomena extensively in the last 30 years of the nineteenth century. By then, there was already a century of debate about the reality of the phenomena, about their therapeutic role and about their moral propriety, especially about women losing control. Many scientists took the line exemplified by the physicist Michael Faraday, who showed with a surreptitiously introduced apparatus in 1853 that table-turning, attributed to spirits, was a response to pressure from the hands of people sitting around the table. Though such investigation encouraged scepticism about spirits, it was of great psychological interest because it pointed to participants acting unconsciously. Moreover, a number of eminent scientists, such as A. R. Wallace and the French physiologist Charles Richet (who won a Nobel prize for work related to the study of immunity), believed what they saw when they attended séances.

In 1882 a group of British intellectuals unhappy with the scientific world view formed the Society for Psychical Research (SPR), which is still active, and led the attempt to examine psychical phenomena objectively.

The first president, Henry Sidgwick, was a Cambridge philosopher, and his wife, Eleanor Sidgwick, who was very active in the society, was the principal of Newnham College, an early college for women in Cambridge. After four years, many of the spiritualist members of the society resigned, unhappy with the SPR for not authorizing the beliefs for the support of which they thought it had been formed. This split expressed the divide between academics, who made objective methods the priority, and ordinary mortals whose priority was hope. The same divide continued to threaten every aspect of ESP research and was visible in the repeated struggles between experimenter and experimental subject for control over the circumstances of the supposed phenomena. Psychological investigators, usually men, wanted to take control so that they could acquire authority to persuade sceptical colleagues and to claim scientific status. The subjects themselves, however, who were often women (though some famous psychics were men), believed that it was openness to control from a spirit realm, not control by scientists, which mattered.

The early psychical researchers, led by Edmund Gurney (1847–1888) and Frederic W. H. Myers in Cambridge, and by James in the other Cambridge, diligently pursued the society's aims to examine scientifically supposed manifestations which defied understanding. They examined the facts, though many scientists thought them naive in their relations with clairvoyants and mediums. All the same, it was the researchers' hope to find facts with a place for human feelings, in contrast to mainstream research which appeared dedicated to excluding feelings. By the 1920s, though the researchers had learned much about deception and illusion, they still came up against the inexplicable. One of the most impressive cases involved the Polish engineer, Stefan Ossowiecki, who exhibited his skills at medical congresses and was examined by the then research officer and later critic of the SPR, Eric Dingwall (1890–1986). Ossowiecki, even under the most stringent conditions, was able to reproduce images and even words securely wrapped inside envelopes. The problem for the researchers was that whatever evidence they put forward for a psychical occurrence, critics were always able to point out the possibility of deception, even when, as with Ossowiecki, no deception was in fact shown. Given that known physical laws did not permit psychical phenomena, and given that deception had been shown or at least suspected in many cases, most scientists assumed no paranormal phenomena were real. It was precisely to escape undecidable evidence about mental powers which led people to become no-nonsense 'behavior men'.

Psychical research, indeed, was an embarrassment to a psychology discipline concerned about its scientific status. Psychologists wanted a respectable science, while spiritualists longed for experience which transcended science as scientists defined it. There was stalemate: scientists were objective but without knowledge of 'more things in heaven and earth . . . than are dreamt of in our philosophy' (in Hamlet's famous line); enthusiasts, by contrast, had knowledge but were not objective.

In this situation, beginning in the late 1920s, J. B. Rhine (1895–1980), with his wife Louise a constant colleague, transformed the methods and institutional standing of psychical research. Yet the results still remained ambiguous. As a young man trained as a botanist but with leanings towards a deeper experience of the human condition, Rhine attended a séance in Boston. Repelled by what he saw as trickery, he became convinced that if there is to be investigation, it should begin with the most elementary claims such as card reading or simple telepathy. Attracted to Duke University in North Carolina by an independent psychical researcher, in 1928 Rhine was offered a post by McDougall, who had himself recently arrived from Harvard. McDougall had long defended animism, a biology recognizing non-material life forces which, he suggested, might explain psychical effects as well as much else. Rhine thus obtained sufficient institutional support to begin systematic work in 'parapsychology', the term he used to describe a field in which rigorous methods would, he hoped, compel scientists to pay attention.

Rhine's ambition was impeccably scientific: to find a regularly reproducible phenomenon to study by objective methods. He devised card-guessing experiments, requiring the subject to identify cards in a specially designed pack (the Zener cards). He then compared the results with calculations of random probability. He introduced many variations, for instance, by asking the subject to guess cards before the pack is shuffled or when aroused by drugs. A succession of high scores in the early trials was crucial for the continuation of the enterprise. But such a succession never recurred. The falling off feature, in which non-random results dropped away over time, left the field without reproducible results – except for the falling-off itself. The Rhines themselves believed that they had finally proved the existence of extra-sensory perception – systematic variation of results with the test conditions was, they thought, particularly impressive – and research could therefore shift, they argued, from seeking proof to seeking understanding. The very dullness and mechanical repetitiveness, the meaninglessness, of their experiments, carried out in a laboratory setting with quantitative results, was, they believed, the key to authority. Yet driving this work was still the kind of hope which had motivated earlier researchers.

Experimental psychologists, sensitive about the enforcement of explicit methods, were prominent sceptics, though the Rhines did enough to secure some authority. *Extra-Sensory Perception* (1934) and *New Frontiers of the Mind* (1937) attracted academic and public attention and popularized the term 'ESP', but private funding was essential for the work and Rhine founded his own journal. Mainstream experimental psychology and parapsychology continued along different paths. After the 1940s, the field had the characteristics of a scientific speciality – an independent research programme, experimental and quantitative methods, a social infrastructure and even a founding father – but it remained marginal. This did not necessarily reflect antagonism from the mainstream but it did reflect indifference: psychologists lacked interest or energy to do research in an area with such a poor record in producing regular results. Who would spend a career gathering data which in the end might prove nothing? There were exceptions, notably Gardner Murphy (1895–1979), a senior figure in the APA and a writer on a wide range of topics. He supervised some work which further showed the lack of consensus. Gertrude Schmeidler demonstrated that the experimental subject's initial attitude towards ESP – people she called 'sheep' express sympathy, 'goats' scepticism – affects the chances of a high score at card-guessing. Her work was conventional in terms of experimental design. But those who supported parapsychology's findings understood her results to reveal further regular patterns in the phenomena, while sceptics assumed the results were biased by previous beliefs. Whatever parapsychologists did, the results were treated as inconclusive, and from the 1940s they gradually withdrew from the effort to integrate with mainstream psychology. This left them socially isolated: research papers appeared only in specialist journals, Rhine's laboratory at Duke University was separate from the psychology department and ESP researchers who moved on to the topics which they really cared about, like survival after death, only further enlarged the social distance.

Rhine sustained his research programme by force of will, and some of those who worked with him regarded him as dictatorial. At the end, several things were apparent. The exhaustive application of scientific method still had not produced reproducible results: rigorous methodology was not sufficient to produce knowledge as ESP's critics simply could not accept the idea of non-material causation. Moreover, many people who were sympathetic to the possibility of ESP became disappointed in the Rhines' work since the rigour of the methods and the triviality of the results detracted from the living realities which gave meaning to psychical life.

For many ordinary people around the world, the inability or unwillingness of qualified psychologists to take psychical experiences seriously indicated the limitations of a scientific approach to human nature. For most psychologists, this showed how much public credulity had still to be overcome by science. There were different evaluations of the authority of personal experience and the authority of experts, not just of facts. ESP research had some temporary institutional success; for example, there were chairs in parapsychology in Freiburg (1954–75), Utrecht (1974–88) and, from 1985, a position in the department of psychology at the University of Edinburgh. At Edinburgh, the incentive and funding came, once again, from outside normal academic channels, in the form of a bequest from the novelist Arthur Koestler. Koestler, who had worked for Stalin's secret service in the 1930s but then become a scourge of scientific materialism, intended to put psychologists on the spot. He left money for research knowing that though institutions would covet the money, they would not want to damage their scientific reputations. Edinburgh found a way to manage this. Then, there were persistent reports of extensive research, supported by the Soviet military, in the period of the early Cold War. In the United States, there was considerable activity in foundations like the Mind Science Center at San Antonio, Texas, where free response techniques tried to capture actual life conditions rather than the forced conditions of the laboratory. Most interesting in the long run, I would think, in many parts of the world, for example, Brazil and Mongolia, belief about non-physical extra-sensory phenomena continued to have an embedded place in culture. It is there, rather than in laboratories, where insight may come.

## READING

The ideal of unified science and reality of disunity in psychology has created a literature on the history and theory of 'schools', and all histories cover this ground. Stimulating, is S. Koch and D. E. Leary, eds, *A Century of Psychology as Science* (New York, 1985), as well as the work of Danziger (see reading for chapter Four). For Britain and Bartlett: A. Costall, 'Why British Psychology Is Not Social: Frederic Bartlett's Promotion of the New Academic Discipline', *Canadian Psychology/Psychologie canadienne*, XXXIII (1992), pp. 633–9; A. Collins, 'The Embodiment of Reconciliation: Order and Change in the Work of Frederic Bartlett', *History of Psychology*, IX (2006), pp. 290–312. For Watson's biography, K. W. Buckley, *Mechanical Man: John Broadus Watson and the Beginnings of Behaviorism* (New York, 1989). For methodology and philosophy of science: B. D. Mackenzie,

*Behaviourism and the Limits of Scientific Method* (London, 1977), and L. D. Smith, *Behaviorism and Logical Positivism: A Reassessment of the Alliance* (Stanford, CA, 1986). Two papers by Franz Samelson began to cut down the idea of a behaviourist revolution to its historical proportions: 'Struggle for Scientific Authority: The Reception of Watson's Behaviorism, 1913–1920', *Journal of the History of the Behavioral Sciences*, XVII (1981), pp. 399–425, and 'Organizing for the Kingdom of Behavior: Academic Battles and Organizational Policies in the Twenties', *Journal of the History of the Behaviorial Sciences*, XXI (1985), pp. 33–47. Ben Harris brought sense to the history of 'Little Albert': 'Whatever Happened to Little Albert?', *American Psychologist*, XXXIV (1979), pp. 151–60.

Daniel Todes has enriched knowledge of Pavlov's life and work: 'Pavlov and the Bolsheviks', *History and Philosophy of the Life Sciences*, XVII (1995), pp. 379–418; *Pavlov's Physiology Factory: Experiment, Interpretation, Laboratory Enterprise* (Baltimore, ML, 2002); and a definitive biography forthcoming, *Ivan Pavlov: A Russian Life in Science*, which will counter the negative picture in Joravsky's study (see reading for chapter Three). Also, T. Rüting, *Pavlov und der Neue Mensch: Diskurse über Disziplinierung in Sowjetrussland* (Munich, 2002).

Descriptive works of the phenomenological tradition by H. Spiegelberg are standard: *The Phenomenological Movement: A Historical Introduction*, 2 vols (The Hague, 1976); *Phenomenology in Psychology and Psychiatry: A Historical Introduction* (Evanston, IL, 1972). For Buytendijk, Dehue (see reading for chapter Three), and J. J. Kockelmans, ed., *Phenomenological Psychology: The Dutch School* (Dordrecht, 1987). There is the standard study of M. G. Ash, *Gestalt Psychology in German Culture, 1890–1967: Holism and the Quest for Objectivity* (Cambridge, 1995). Also for holistic thought, A. Harrington, *Reenchanted Science: Holism in German Culture from Wilhelm II to Hitler* (Princeton, NJ, 1996). On the gestalt psychologists in North America: M. G. Ash, 'Gestalt Psychology: Origins in Germany and Reception in the United States', in *Points of View in the Modern History of Psychology*, ed. C. E. Buxton (Orlando, FL, 1985), pp. 295–344; M. M. Sokal, 'The Gestalt Psychologists in Behaviorist America', *American Historical Review*, LXXXIX (1984), pp. 1240–63.

There are two excellent studies of the history of parapsychology, one general, the other on the Rhines: J. Beloff, *Parapsychology: A Concise History* (London, 1993); S. H. Mauskopf and M. R. McVaugh, *The Elusive Science: Origins of Experimental Psychical Research* (Baltimore, ML, 1980). There is a good study of Murphy's work: K. Pandora, *Rebels within the Ranks: Psychologists' Critique of Scientific Authority and Democratic Realities in New Deal*

*America* (Cambridge, 1997). For the international dimension, E. R. Valentine, ed., 'Relations between Psychical Research and Academic Psychology in Europe, the USA and Japan', special issue, *History of the Human Sciences*, xxv/2 (2012).

CHAPTER SIX

# Unconscious Mind

> Always the same reversal: what the world takes for 'objective,' I regard as factitious; and what the world regards as madness, illusion, error, I take for truth. It is in the deepest part of the lure that the sensation of truth comes to rest.
>
> Roland Barthes

## SITUATING THE HISTORY

Sigmund Freud (1856–1939) rested his claim to immortality on a by now clichéd trope: human reason, which once thought itself the centre of the universe, first found itself on a lonely planet, then learned it was descended from the monkeys and now, through Freud's own work, has understood reason not to be reason at all. It was, of course, also a story about heroism, with Copernicus, then Darwin and now Freud pressing towards truth in the face of superstition and weakness of will. Freud writ large the message that, as a scientist, he fulfilled the ancient imperative, 'Know thyself'.

Psychoanalysis and other so-called depth psychologies did not just form a 'school' alongside behaviourism, Pavlovianism or whatever. Their long-term significance lay, I think, in fostering alternative forms of understanding, more akin to literature than to biology. Yet Freud, C. G. Jung (1875–1961), Jacques Lacan (1901–1981) and others claimed the standing of science for their work, even if many critics argued exactly the absence of this standing vitiated it. Once again, there are interesting differences on the matter of what psychology *is* to sort through. It is remarkable that though Freud is probably the name best known to the public in the whole of modern psychology, academic psychologists commonly ignore, or even denigrate, his ideas.

I use Freud's word 'psychoanalysis' to describe his theories and practice and their many derivatives. Jung's work had independent roots and formed an independent project, even if Jung was for a few years close to Freud, and I therefore distinguish it by Jung's term, 'analytical psychology'. Reference to unconscious processes was ubiquitous in the nineteenth century, but there was a difference between talk in the romantic manner, about unconscious psychic powers, statements about events going on in the nervous accompaniments of mind (like Helmholtz's 'unconscious inference') but not accompanied by consciousness and theories, like Freud's and Jung's, which specifically referred to an unconscious mind. Freud and Jung originated psychotherapies, though there was a long history of psychological approaches to therapy (such as in spiritual healing and in mesmeric séances) independent of them. Taken together, all these therapies had a place in treating the mentally ill, and in this context they constituted 'dynamic' (as opposed to organic or biological) psychiatry.

All forms of belief about a hidden realm have something of the numinous about them: they draw on a deep well of sensibility, crossing the centuries and crossing cultures, about mystery beyond everyday life. Theories of the unconscious mind have shared this numinosity and, as a result, their story lies as much with cultural history as with the history of psychology. Freud, for instance, was attached to the archaeological metaphor, which located the secrets of the unconscious in early time as well as in psychic space; and there was a public as eager to read about digging up ancient treasure as about the uncovering of unconscious layers of the psyche. Debate inevitably opened out into discussion of psychology and religion – as Freud himself made clear by asserting his claim to stand on the shoulders of Copernicus and Darwin. What he saw, he upheld, was reason demonstrating the limits of reason. There was an audience for the view that reason comes up against obstacles embedded in being human. This would not perhaps have confounded a religious age knowledgeable of the falling away of men and women from God, but it did give rise to anguish in a self-described enlightened age. Individual distress accompanied political failure. Even when people achieved unprecedented privilege and material well-being, there was unhappiness, depression, aimlessness and fear. What stories could people tell themselves to understand this? One narrative used the language of the unconscious, and it became an international language of psychological society. When, late in the twentieth century, psychologies of the unconscious came under fierce attack from hereditarian and biological theories in psychology, the latter, too, promoted beliefs about human nature which demonstrated limits to reason, though attributing them to genetic inheritance not unconscious forces.

Freud, the son of Jewish migrants from Moravia to Vienna and in the last year of his life a refugee from Hitler, was a figure of the Enlightenment. He sought a science of human nature as a guide to how to live, and as an enemy of superstition and an emancipator of people from ignorance about the intimate world, his reputation is inextricably linked to sexual liberation. At the same time, he wrote as a prophet, warning the world about the power of the unconscious, an irrationality deeper than reason: 'The unconscious is the true psychical reality.' Freud's message was pessimistic, yet the source of this pessimism was also the source of an optimism that enlightened reason, which, as Freud saw it, must act in the modern world without benefit of spiritual grace, can improve the human condition. He had followers who looked to psychoanalysis to redeem individuals, even the world, and for the means to deal with 'eternal' questions of birth, death, love, hate, rage and loss. He also developed *talk* to manage the particularities of individual suffering – phobia, anxiety, loneliness, obsession, emptiness and inadequacy of every kind.

In an unguarded moment, Freud asserted: 'I am actually not at all a man of science, not an observer, not an experimenter, not a thinker. I am by temperament nothing but a conquistador – an adventurer . . . with all the curiosity, daring, and tenacity characteristic of a man of this sort.' He identified himself as a hero who quests for the light of truth though aware that the journey may end in darkness. He was a brilliant intellect of liberal scope who was, of course, at the same time a flawed individual. He was a man of reason and culture who was nevertheless sometimes overwhelmed by petty spite and ambition, and he deceived himself as well as others about the origins and development of his work. As scholarship has made clear, he fabricated an account of the origins of his thought and used this to underpin the authority of what he had to say. Once his own account is questioned, the distinctiveness of Freud's work and of psychoanalysis as a movement begins to dissolve. More deeply and personally than he perhaps knew, he was in trouble: if the unconscious, irrational mind structures language and action, on what basis can we accept as truthful anything that we say or do? Freud straddled the nineteenth and twentieth centuries: educated in the certainties of nineteenth-century natural science, he developed a psychology of the uncertain inner life characteristic of contemporary times. He is a figurehead for loss of innocence.

Freud's stated purpose, however, was to contribute to science. He made a series of claims for which he became famous: the existence of a dynamic, continuously active unconscious mind; the sexual theory of the neuroses and, by extension, the sexual theory of motivation in general; the interpretation of dreams as wish-fulfilling fantasies; the Oedipus complex as the basis of individual character and moral culture.

Following his education and research in bio-medicine, Freud created psychoanalysis between 1885 and 1900 while working mostly in private practice as a specialist in nervous diseases. He added dream analysis. His ambition was to create a general psychology, yet the empirical authority for his science was a small number of cases, not least the case of Sigmund

Freud, in a middle-class consulting room. He saw how his writing had much in common with the short story, and others have appreciated the comparison between his methods and those of Sherlock Holmes. He tried to persuade readers that his stories made human actions, even the most trivial, like slips of the tongue, intelligible. His manner of working can be compared with contemporary French clinical studies in psychology. Indeed, French psychologists thought that their colleague, Janet, had priority or even was the source for Freud's understanding of the dynamic unconscious. In his studies of 'Léonie', Janet had posited the existence of a 'subconscious' in order to explain her dramatic powers, like her three personalities and apparent ability to fall into hypnotic sleep at a distance. From 1902, for 30 years, he lectured on the normal and abnormal mind, sharply critical of Freud's sexual theory of the neuroses. But while Janet's early work attracted international attention, his later writings did not, at least to the same degree. His institutional position gave him few students, while his presence in the Collège de France perhaps helped block interest in psychoanalysis in French culture before the 1930s.

Later, from about 1950 to 1980, when psychoanalysis excited interest among trained psychologists in the United States and had a large influence in American psychiatry, psychologists studied Freud's ideas by experimental methods. The results were negative. The distance between Freud's ideas and anything which experiments showed appeared to confirm, his protestations notwithstanding, that psychoanalysis was simply not a science. Then, historians began to examine the sources of Freud's own case studies and to demonstrate the extent to which he indeed wrote stories, in the sense in which popular language uses the term. From about 1980 it became common to denigrate his views, though psychoanalytic training and practice continued to attract a small, intense following, more interested in therapy and culture than natural science.

For all this, Freud was scientifically and medically trained in anatomy and physiology to the highest standards. It was enthusiasm for science and ambition to make an impact as a scientist which led him from clinical detail to general laws, and it was only the need to earn a decent living which forced him out of the laboratory and into medical practice. This background strongly coloured his new psychology: he held to determinism and belief that everything has a specific cause; and he conceived of mental structures and functions in terms of physiological models (for example, of energy and a bodily economy). He discussed individual development and historical culture in the light of evolutionary theory. Many scientists accordingly thought it right to assess psychoanalysis as a natural science and then, finding that it did not meet the standards, to dismiss it. These critics then attributed Freud's fame to public gullibility, passing intellectual fashion or the seductions of persuasive writing.

None of this begins to explain Freud's historical importance. When Freud wrote about his patients, the unconscious and sexuality, he was aware – sometimes painfully so, since he ardently wished to be original – that he was rephrasing insights embedded in ancient myth, in literature and in critical thought. Freud was well read and a great stylist in the German language and received the Goethe Prize in Frankfurt in 1930. He collected antiquities, and psychoanalysis and archaeology shared a vision of recovery by the excavation of the covering layers of history. He brought all of this as well as his training in natural science to bear on his patients' symptoms and a general theory of the psyche. A deep ambivalence persisted. He imagined the unconscious as a mental realm in parallel with brain events (and hence structured like brain events by energy dynamics), which, one day, would be known in biological terms. And, simultaneously, he turned aside from biology to develop an account of a true psychic unconscious, a realm of mind *sui generis*, with its own structure and laws.

Some people have therefore read and judged Freud as a natural scientist, while others have read and judged him as an intellectual of a different kind, one who searched for a systematic, consistent and objective but not natural science understanding of the psyche. The latter reading judges his contribution as interpretive work of the kind to be found in literature, or indeed in the kind of history I am now writing. This is supportive of an assessment of his work as a practical response to individual suffering and a key source for modern moral philosophy.

These issues have involved a debate about Freud's language and its translation from German into English. The *Standard Edition* of his psychological work appeared in English between 1953 and 1974 under the general editorship of James Strachey (the brother of the literary writer, Lytton Strachey). In many respects it was a monument to careful scholarship, it transformed access to Freud for non-German-language readers and it was substantially responsible for the enormous amount of English-language commentary on Freud. The edition had its critics, however, who stated that Strachey and his translators, seeking acceptance among medical psychiatrists, deferred to natural science and found words with appropriate scientific coloration in English where Freud used German words with humanistic or even spiritual connotations. '*Seele*' was rendered in English as 'mind', not as 'soul', and '*das Ich*' was rendered as 'the ego', not as 'the I'. Critics also pointed out that when Freud discussed psychic energy, in

English this sounded mechanistic, whereas he may have had in mind a philosophy of will, like that of Schopenhauer or Nietzsche. There was therefore a new English translation, under the editorship of Adam Phillips.

All this argument was not restrained and abstract but passionate and concrete: it stirred up emotionally unsettling powers within the psyche and sexual life. The disruptive potential of psychological quests had surfaced in earlier controversies, for example, with Mesmer, and appeared again later, for instance, in the late twentieth-century furore over the balance of memory and fantasy in reported child abuse. When Freud elaborated dream analysis – talking with the inner other – he touched on ancient interests in altered states of consciousness, psychic healing and the power of confession. His readers knew the romantic imagery of the hidden powers of the soul, and they were notably preoccupied with sex – the repression of public discourse mirroring private obsession. Indeed, when Freud began to publish, sexology was already an established medical domain. History should do much more than focus yet again on Freud; all the same, Freud's life and work is our point of reference.

## PSYCHOANALYSIS: THE EARLY YEARS

Doctors, like Freud, who specialized in the borderland of mind and body faced an utterly confusing hotchpotch of hysterical and nervous symptoms caused by unknown combinations of mental and physical factors. Though the symptoms and focus of attention have shifted, the situation is easy to imagine: depression, trauma, autism, addiction and the placebo effect offer comparable challenges. In Freud's time, there was little agreement whether many symptoms were real, put on or imagined, though there were legal and financial, as well as personal, implications (for example, with 'railway spine' following accidents). There was a late nineteenth-century outbreak of 'neurasthenia', which was a recently coined term for lassitude, weakness, loss of will, with such symptoms as sensitivity to light or noise and headaches. Hysterical symptoms were often dramatic, with partial paralysis, hallucinations, compulsive actions and prostrating anxiety. Men and women suffered alike, though language was highly gendered.

The medical response to this borderland world was as polyvalent as the symptoms themselves. Many medical authorities, like Freud's superior at the psychiatry department of the Vienna general hospital, Theodor Meynert, took a physicalist approach, believing that all disorders, if genuine, are physical in nature, even when the symptoms are functional, that is, when the symptoms appear as disordered activity not visibly due to organic damage. To attract patients, Freud needed to make sense of and to offer cures for the bewildering disorders which came into his consulting room. He was willing to experiment: he used hypnosis and cocaine, for example, as well as more conventional rest cures and electrotherapy. He also had a driving intellectual ambition as a scientist to reveal causal order beneath the surface confusion.

When Freud started out as a doctor, Josef Breuer (1842–1925), a well-established Viennese physician, helped him emotionally, intellectually and by referring patients. Among Breuer's patients was Bertha Pappenheim, celebrated as the first psychoanalytic subject, 'Anna O.', whose case Breuer and Freud wrote up in their joint *Studien über Hysterie* (1895; *Studies on Hysteria*). She exhibited a kind of self-induced hypnosis in her disturbed condition, and Breuer found that if he then encouraged her to talk about memories this had some therapeutic effect. This creative patient called the talk 'chimney-sweeping', while the doctors referred to 'catharsis'. If the probing of historians has shown the fabricated nature of Freud's version of what went on, he nevertheless used this (and similar) material to elaborate psychoanalysis. According to Freud, it dawned on Breuer that his patient renewed her symptoms because she was in love with her doctor, and Breuer therefore broke off the relationship. Freud interpreted symptoms as symbols of events within a hidden unconscious, and he utilized the patient's feelings for her doctor as emotional energy for a therapeutic relationship.

Freud gradually shifted from physical to psychological explanation, and here his experience of hypnotism was decisive. He obtained a grant to study in Paris in the winter of 1885–6, with a base in the Salpêtrière hospital, where Charcot was then at the height of his fame as a neurologist. At this time, in Nancy, the regional centre of Lorraine, a local doctor, A. A. Liebéault (1823–1904), followed by an academic physician, Hippolyte Bernheim (1840–1919), used hypnotism regularly as a therapy, claiming it had healing power for everyday illnesses. With Charcot's interest, however, hypnotism moved in from the medical margins and ceased to provoke censure from the medical establishment. The Salpêtrière, where Charcot worked, was more like a small town than a hospital and it provided Charcot with a vast array of nervous disorders to study. He imposed descriptions and a classification, confirming neurology's status as a medical speciality with a theatrical flair that made him a celebrity. He boldly included hysteria within his remit, persuaded his colleagues that it was a real disorder and then used hypnotism as a technique to reproduce and study its symptoms. The Nancy physicians, meanwhile, using hypnosis as a therapy, attacked Charcot's whole approach, fuelling a dispute which gained a large audience.

Charcot, like Meynert, looked for physical explanations. He assumed that hysteria had a physiological and inherited basis, though he also perceived the way hysterical patients exhibited symptoms which followed ideas and not anatomy: hysterical paralysis, for example, affected parts used in expressive movement rather than parts linked by the same nerves. For Charcot, this was of interest; for Freud, it became the key evidence for symptoms as the symbolic expression of something in the mind of which the patient is unaware. In Breuer and Freud's famous phrase: 'Hysterics suffer mainly from reminiscences.'

Freud expanded his ideas after he returned from Paris and a later visit to Nancy. But he did not find hypnotic techniques very satisfactory and instead devised what he called 'free association' and then used dream analysis in order to gain access to hidden memory. He built on two fundamental principles: first, visible conscious psychic life is not itself reality but symbolically represents unconscious reality; second, determinism – everything has a cause which is part of a regular pattern. Both principles rendered the small change of daily life, like jokes, evidence for a general psychology. This is one important reason why Freud achieved a public audience. Freud then introduced the couch into the consulting room and obliged his patient to report 'whatever comes into his head . . . not being misled, for instance, into suppressing an idea because it strikes him as unimportant or irrelevant or because it seems to him meaningless.' Psychoanalysis became a *method* to gain access to the unconscious, a *theory* about what drives the unconscious – 'the mechanism of defence' – causing repression, and a *therapeutic practice*, 'the transference', which draws out the emotional energies of the unconscious through a psychological relationship between analyst and patient. With all this, Freud never disowned an implicitly philosophical ambition, evident, for instance, in passages of his writing where he recognized mystery at the heart of human existence.

Going beyond the analysis of individual repression and symptom formation, Freud continuously returned to metapsychology, to a theory of the general structure and nature of psychic life. He posited the existence of libido, psychic energy, a life force characteristically but not uniquely expressed in sexuality. It is libido's nature, he argued, to pursue pleasure ('the pleasure principle'), but the conditions of social life mediated by the family severely restrict its scope ('the reality principle'). Child development turns on these two principles confronting each other in the (male) child's Oedipus complex and its (normal) resolution.

After the publication of the book on hysteria, when he pressured Breuer into co-authorship, Freud brought out his major work on *Die*

*Traumdeutung* (1899–1900; *The Interpretation of Dreams*), with its theoretical chapter Seven, and other books on the psychology of everyday life, jokes and sexuality, along with case studies. These works gave psychoanalysis a public life. A small group began to meet every Wednesday evening in Freud's apartment at Berggasse 19 (which became a Freud museum in the 1980s) and formed the Vienna psychoanalytic society. His publications were not ignored, as Freud and other early analysts liked to claim in order to enhance their self-image as lonely explorers, but neither was the work recognized as a new science, as Freud thought it deserved to be. *The Interpretation of Dreams* inspired a small number of intellectuals, who were mostly not physicians, and who were all also Jewish, who formed the early Vienna group. In the wider world, it was admired by the Swiss psychiatrist Jung, and his younger colleague, Ludwig Binswanger, and their visit to Freud in 1907 led to the foundation of a psychoanalytic group in a medical university context in Zurich. This marked the movement's expansion onto the international scene and away from its Jewish context in Vienna. With Freud's and Jung's visit to the United States in 1909, and the foundation of the International Psychoanalytic Association in 1911, it began to have a public and medical presence.

Freud followed the established method in clinical science when he generalized from particular cases. He did not, however, put forward only a medical theory of the neuroses but used his cases (including himself) to create a general psychology, with which in turn he interpreted morality, culture and religion. Jung was to do the same. It was a bravura performance. Freud combined stories about a governess, 'Miss Lucy R.', who could not escape the smell of burnt pudding, with references to sex and questions about the meaning of life. This gained Freud an audience which no number of studies on nonsense syllables or rats in mazes was able to do.

The popular reputation of Freud's psychology, from the beginning, was that it emphasized sex at the expense of everything else. People have been inclined to attribute changing mores to a founding father, though changes in attitudes to sexuality were certainly not all due to Freud. The term 'sexuality' itself became current just at the time his work appeared, and a growing number of physicians, teachers, anthropologists, writers and moralists were publishing in the area. They included Freud's psychiatric colleague Albert Moll, who emphasized childhood sexuality, and another colleague, Richard von Krafft-Ebing, who classified sexual abnormalities and coined the terms 'sadism' and 'masochism'. A contemporary writer, Arthur Schnitzler, portrayed bourgeois marriage as a tragedy which frustrates sex inside marriage and makes happiness in sex impossible outside marriage. Gustav Klimt, the leader of the Secession, a group of young artists who left the official art academy, brought the direct expression of eroticism into painting. Richard Strauss, the composer, in 1908, staged his opera *Salome*, with a portrayal of violent sexuality. This was just Vienna. In Britain, Havelock Ellis (1859–1939) published his encyclopaedic *Studies in the Psychology of Sex* (7 vols, 1897–1928); in the United States, Hall focused attention on adolescence, and it was Hall who brought Freud and Jung across the Atlantic for a series of lectures.

Breuer had observed that 'the great majority of severe neuroses in women have their origin in the marriage bed.' As Schnitzler's stories and Freud's patients confirmed, this was no idle joke. Freud's sexual theory developed when he tried to understand the material his patients repressed. On the couch, asked to speak freely, his patients resisted, refusing or being unable to follow the associations of their thoughts. Freud traced symptoms to childhood and found sexual content. For a while, he thought this content derived from a traumatic experience, and he concluded that his patients had experienced childhood seduction, abuse, causing pain to remain in buried memory until adolescent or adult experience released the energy as symptoms. Then, Freud began to examine his own sexuality, as well as to reconsider his patients' reports, and he became convinced that the memories were in reality memories of fantasies. This implied that young children possess sexual feelings which are expressed and then repressed in their relationships with parents. Freud constructed a general theory of sexual development, but he rejected the accusation of bias towards sex. When he pointed to sexual forces behind ordinary life, he claimed, like any scientist, only to trace what we observe to its causal roots.

At this stage, Freud presupposed the existence of instinctual energy but gave little attention to what it is, rather than to its effect, repression. His notion of libido was quite wide, involving desire for pleasure rather than sex in a narrow sense. He adopted speculative notions common in nineteenth-century biology, notably belief in the inheritance of acquired characteristics and collective psychic ancestry. Freud was overwhelmingly concerned with male sexuality, even though a substantial number of his patients were women. He never seriously questioned a male-oriented norm of sexuality, a weakness later painfully exposed. Finally, his concentration on sexuality was not primarily a call for its liberation, though he did hope reason would replace prejudice. He remained a bourgeois man of his time and he demanded sexual containment as a condition of civilized life.

## THE PSYCHOANALYTIC MOVEMENT

Freud had insatiable intellectual curiosity and ambition. Many of those he attracted in the early years, like Alfred Adler (1870–1937), Sándor Ferenczi (1873–1933) and Lou Andreas-Salomé (1861–1937), looked for something more like a guide to life than a psychology. For Helene Deutsch (1884–1982), before she modified her views on femininity and motherhood in the 1930s, Freud was a guiding light in dark times. If psychoanalysis began as a medical practice to heal individuals, many of its followers hoped it would become a movement to heal the world. It made an inestimable contribution, especially in the 1920s, to forming a social world in which people reflected on their lives in psychological terms. Whatever the rhetoric about the scientific character of Freud's work, he and his followers perceived it to have a mission. But this was true for many movements in psychology and in science generally. Distinctively, psychoanalysis had at its centre a theory why the world *resists* enlightenment. This was a strength because, like religion, it offered a way to come to terms with human self-destructiveness. But it also had a negative consequence, since it encouraged Freud and his closest followers to form a sect dogmatically loyal to his supposed insights, explaining away criticism as resistance.

There was a rapid growth of interest in psychoanalysis just before 1914 in the English and German-speaking worlds. Freud extended his reach to culture and civilization, and he published on the spread of nervous illness, Leonardo da Vinci and the origin of taboos. The analysts consolidated themselves into a movement and Jung, at Freud's insistence, became the first president of the International Psychoanalytic Association in 1910. This was immediately followed by a schism: Adler went his own way in 1911 and Jung left in 1913. These defections, from Freud's point of view, led to the establishment of 'the committee', an inner circle of followers, to each of whom Freud presented an intaglio ring.

Adler was one of the first members of Freud's group in 1902, and his departure was a blow to Freud. Adler, however, had always had a commitment to social medicine in the welfare of humanity. He did not fully accept Freud's sexual theory of the neuroses and recognized the existence of an autonomous aggressive drive, which he linked to functional relations between bodily organs and to family and social relationships – as the middle brother of three, he had a special interest in sibling rivalry. These differences became public after the break with Freud, when Adler published *Über den nervösen Charakter* (1912; *The Nervous Character*), and especially after the war. He believed that psychic forces work in the interest of the unique individual, as if each individual has a distinct goal. When a person deviates from this goal, he thought, this produces a neurosis. Adler also developed the notion of the feeling of inferiority as a significant factor, a notion which began life as an orthodox medical theory about how a bodily organ may have a congenital weakness and hence be a likely site of pathology. He translated this into a psychological theory, finding feelings of inferiority to be central in neuroses and ordinary motives (for example, to exert power over others). Neurotic people, he claimed, have an 'inferiority complex'. With his own feelings in mind, Adler linked strong feelings of inferiority to sibling rivalry; he also attributed such feelings to social position – to employees' and women's lack of power.

During the 1920s, Adler developed what he called 'individual psychology' and tried to educate people in community feeling through understanding character, neurosis and aim in life in interaction with the social environment. For Freud, this was just yet more woolly thinking caused by resistance to unpleasant truths about the libido. The postwar educational reforms in Vienna, however, provided Adler with a setting in which to propagate his ideas, especially after he became a professor at the pedagogical institute in 1924. He set up child guidance clinics and contributed to adult education. He later emigrated to the U.S. as he saw a future for his ideas there. He lectured and wrote in an accessible way about the links between personal mastery and early experience, and this was something many ordinary people wanted to know about. His book, *Menschenkenntnis* (1927; translated as *Understanding Human Nature*), conveyed this practical concern in its title. Indeed, the very practicality of Adler's notions resulted in them being absorbed into North American culture and into ego psychology without much awareness of their source.

Meanwhile, there was war, and Freud, a loyal Austrian with two sons in the army, turned in on his own reflective resources. Though long anticipated, and in many quarters ardently desired, war shattered emotions as well as bodies. Observers portrayed war and the revolutionary uprisings which followed as the outbreak of the animal, the instinctual and the irrational in human nature, stripping away the veneer of progress. Such language resonated with Christian belief in original sin. Freud responded with a sombre portrait of the unleashed unconscious.

Freud attributed social relations to instincts worked out in the family's socialization of each child. He located the origin of civilization in the rebellion of the sons against the original father, the father's murder and the subsequent suppression of fratricide, the sons' attempt to kill each other in competition for the father's wives, by the internalization of guilt and by the institution of the incest taboo. The primal horde of mankind thus acquired a social order and the basis for law when it repressed the sons' desire to sleep with the mother after it had killed the father. Elements of this ancestry, Freud thought, persist as an inherited psychic structure in the modern mind. This mythology had a background in late nineteenth-century anthropological speculation, but in the new century there was a turn away from such work in favour of field studies and conceptual clarity. Freud's anthropology remained the least influential part of his writing.

Freud tried to understand the collective psychic forces which, he thought, found expression in the war. He rejected belief in the existence of an autonomous social instinct and instead attributed social solidarity to sublimated sexuality, to 'Eros, which holds together everything in the world'. When societies collapse, he believed, as German, Austrian, Hungarian and Russian society did at the end of the First World War, repressed feelings of envy and aggression break out. He delineated more precisely what he meant by instinct and what he considered the cost in individual unhappiness, guilt and anxiety of the effort to maintain civilization. In a series of short books, he clarified a structural theory of the psyche, describing the id, the ego and the super-ego. The unconscious id is the instincts or energizing power of the mind. The ego, both conscious and unconscious, arises as a precarious and always tense integration of the instincts and the social world. The material realities and the values of the social world, imposed on the child through the mediation of parents, gives rise to the super-ego, which utilizes the energy of the instincts and turns against the ego, sometimes as destructive feelings of guilt, sometimes, through sublimation or redirection of its energies, as a source of creative power. Contemplation of psychic forces led Freud to conclude that they are life affirming and the source of sexual libido, Eros, but also destructive and life denying. Four influences were at work here. First, war and social upheaval gave Freud and his contemporaries experience of the human capacity for sheer barbarity. Second, Freud interpreted the pleasure principle as the expression of an elemental desire to restore equilibrium, and he thought of the ultimate equilibrium as the state of death – 'the aim of all life is death', as he stated, with words from Schopenhauer. In addition, Freud was puzzled in his clinical work by the tenacity of certain disorders (which will not go away even though they damage the person), like obsessional anxiety. He hoped to explain this kind of self-destructiveness with an enriched conception of the instincts. Last, Freud himself faced death: he feared for his sons in the war and for himself as he aged and suffered from cancer, and he lost a dearly loved daughter, Sophie, and a grandson.

Freud reached into myth as well as into science to write his books and the result, some thought, resembled a religion. Freud, however, in *Die Zukunft einer Illusion* (1927; *The Future of an Illusion*), specifically argued that religion is wish-fulfilling and essentially childish in nature. In *Das Unbehagen in der Kultur* (1930; translated as *Civilization and Its Discontents*), he described the tenuous condition of civilization, which is always threatened by the repressed and sublimated powers that made it possible in the first place. We have a stark choice: to be civilized and unhappy or to destroy ourselves. The one hope, to which Freud was committed, is reason, and the capacity to establish systematic, objective knowledge – science – which may enable us to ameliorate our condition. In a memorable and oft-quoted admonition to an imagined patient, Freud declared: 'But you will be able to convince yourself that much will be gained if we succeed in transforming your hysterical misery into common unhappiness.'

There was a personal dimension to this view of life. In 1923, Freud was operated on for cancer of the jaw, and he lived thereafter with a prosthesis and in daily pain. He refused analgesics and continued to smoke cigars in a conscious decision to preserve his alert mind, to pursue enlightened dignity though faced by the prospect of death. After 1933, he witnessed the destruction of his books in Germany, where the Nazis reviled his work as Jewish science, but he refused to emigrate. With the *Anschluss* he finally left Vienna, after payments to the Nazis, and he came to London as a refugee for the last year of his life. Freud himself did not comment much on politics, but other writers found in his work the terms with which to describe unreason and hate.

Two priorities for Freudians were deepening the therapeutic process and training future analysts. In the tradition of the hypnotic healers Freud had visited in Nancy, psychoanalytic therapy involved a *psychological* relationship between therapist and patient. The analyst became the object of the emotions, whether of love or hate, of the patient's repressed unconscious, and the analyst then sought to redirect those emotions onto a more appropriate person, occupation or mental activity. This placed a heavy burden on the analyst, who was required to be objective and unemotional whilst the object of the strongest feelings. Training therefore required analysis of the future analyst, to make it possible for her or him to achieve objective self-understanding and hence detachment. Freud, in a story which mythologized his achievement, supposedly began the process by self-analysis. Further, before his followers institutionalized formal training, Freud had analysed his own daughter, his youngest child, Anna, which was an extraordinary act given the nature of the repressed contents his theory presupposed. A member of 'the committee', Karl Abraham (1877–1925), a physician, had the abilities needed of an organizer, and he turned Berlin into a major centre for training alongside Vienna. By 1920, Abraham and his colleagues had established a polyclinic which brought treatment and research together in a single centre. During the next decade, several Viennese and Hungarian analysts moved to Berlin, and Anna Freud taught there in 1929. Karen Horney (1885–1952), Melanie Klein (1882–1960) and Deutsch all practised in the city, though Klein left for London in 1926 in reaction against opposition to her methods from her Berlin colleagues. At the same time, people still made the journey to Vienna in the hope of training with the master.

The issue which proved most divisive was whether to require medical training for anyone who wished to become an analyst. Freud, though himself a doctor, resisted this, and indeed many of his earliest and closest supporters, like Anna Freud (1895–1982) herself, were not medically qualified. This created problems with the medical profession, the occupation from which Freud expected recognition. It raised, as mesmerism had done earlier and clinical psychology was to do later, the vexed issue of the social and legal authority of healing by those who are not doctors. In the United States, doctor-analysts imposed a medical requirement, though in spite of this, or perhaps because of it, psychoanalysis saw there its greatest institutional expansion, between about 1945 and 1975, thus achieving a pre-eminent position in psychiatry. Organizational decisions proved to be inseparable from theoretical disputes, and disagreements continually surfaced, often with an emotional intensity which was a source of wonder to outsiders and one reason for the cliché that psychoanalysis is a religion.

It is more helpful to understand that psychoanalysis was just one of a number of psychotherapies with their roots in nineteenth-century alternatives to physical medicine and in ancient traditions of healing and spiritual inquiry. The emergence of differences in the Freudian camp was a natural continuation of a rich culture rather than merely narrow sectarian dispute (though, of course, it was that too).

In Vienna and Berlin, young and sometimes radical analysts worked uncomfortably alongside a more conservative older generation. The radicals constructed a significant political argument, a social psychology, linking personal repression and misery to the conditions which capitalism imposed on labour. They also struggled with the practical problem of making analysis available to those who could not pay. Freud's close colleague Ferenczi held a university position in the short-lived communist government in Budapest in 1919. His practice during the 1920s caused increasing disquiet among other analysts because he questioned the taboo on the analyst showing any emotional reaction to the person on the couch. The majority of analysts thought neutrality was the way to defend objectivity and propriety and, hence, was crucial for their public image. Ferenczi, however, called for an analytic responsiveness dangerously close to emotional involvement. Most explosively, Wilhelm Reich (1897–1957), who, though young, had an important position in analytic training seminars in Vienna during the 1920s, turned to analysis as a tool of sexual liberation. He argued that capitalist interests literally become embodied in workers; it can be seen in the repressed posture and personality of workers who accept poor material conditions and feel guilty at the thought of bodily pleasure. Reich established clinics in working-class districts of Vienna, and later Berlin, to restore sexual desire to workers and hence desire to control their own lives. Reich and Ferenczi, in their different ways, thus tilted the balance – always precarious – which psychoanalysts maintained between the pursuit of science and the pursuit of emotional liberation.

In France, psychoanalysis was in a weak position. The rapid collapse of Charcot's reputation after his death in 1893 left a lasting suspicion about hypnotism and suggestion. His former protégé, the neurologist Joseph Babinski (1857–1932), insisted that suggestion can cause and cure all psychological symptoms – an argument which rendered psychological symptoms of little medical interest and marginalized Freud's response to the neuroses. Janet's prominence as the francophone proponent of the psychology of unconscious forces also helped shut out Freud. Paris intellectual culture, in any case, was not inclined to look beyond itself and exhibited a degree of haughty isolation from the wider world. Nevertheless, a small but intense analytic community established itself during the interwar years. There was a continuing interest in hysteria, focused on 'shell-shocked' soldiers and the new woman, the woman who tried to escape her traditional roles. Most dramatically, the Surrealists celebrated the unconscious in art, theatre and writing, and they assigned to the unconscious positive attributes precisely where polite society found negative ones.

In 1936, Lacan, a relatively young analyst, influenced by Surrealism, read a paper at the annual international psychoanalytic congress, and this signalled a new direction. He drew attention to the baby's response to the image in the mirror and identified this response as the distinguishing feature of the human, signalling the human capacity for psychic life unconsciously structured in terms of images ('imagos', in psychoanalytic jargon), and as the origin of 'the I'. In further talks, he emphatically named

Freud as the man who had made psychology a science, the science which has the symbolic content of the unconscious, structured like a language, as its subject-matter: 'it is in the experience inaugurated by psychoanalysis that we can grasp by what oblique imaginary means the symbolic takes hold in even the deepest recesses of the human organism.' For two decades, in the 1950s and 1960s, Lacan had a huge impact: he lectured to audiences spellbound by his way with words, and, at one time, about 1,000 people, including leading intellectuals, jammed his classes.

Lacan's starting point was antagonism to psychiatry, which treated mental illness as psychologically meaningless, and to the path of psychology, in both behaviourist and Freudian manifestations, in the United States. After 1945, in his own distinctive way indebted to the philosopher Hegel, he developed the notion of unbounded '*désir*' (longing, desire, wish) which seeks confirmation of its being by internalizing the mirror image and the image of others. The psyche therefore has a symbolic order and does not contain a substantial self or a representation of what common opinion calls 'the real world'. The symbolic order, all the same, Lacan argued, taking up the work of the Swiss linguist Ferdinand de Saussure, has a determinate structure in the form of a language, and hence it is the subject-matter of consistent and grounded knowledge, that is, of science. The conscious encounter with the unconscious is an encounter with 'the other', and psychic life, as when a man faces his *Doppelgänger*, is alien, uncanny, unsettling. The unconscious is a realm of language which constantly escapes the attempt to control it by a conscious subject. The technique of psychoanalysis, Lacan claimed, is the route to the objective understanding of the unconscious; for instance, analysts recognize the symbol of the phallus as the most powerful signifier of *désir*. (Whether the privilege thus accorded to the phallus marginalized women later became the subject of heated controversy.)

For Lacan, play with language, the kind of play which ruled his own style, is the means by which the unconscious can speak. Any claim to grasp the unconscious, the other, in plain speech, is a delusion. Yet he demanded his colleagues return to Freud's texts as the basis for scientific psychology, and, in the 1960s, he turned to mathematical formalisms to give his theory more rigorous or scientific expression (though this side of his work had little impact). Other claims about psychology he rejected outright. In 1953, his refusal to conform to the conventions of psychoanalytic practice – the brevity of his therapeutic sessions became legendary – led the Société Française de Psychanalyse to split from the Société Psychanalytique de Paris. Later, especially after the political events of 1968, in which Lacan acquired the image of an anarchist hero, he emphasized even more the way in which the unconscious, through language, always goes beyond what we can know or grasp and spills over in contradictions, play, seductions, irony and desires different from what our conscious, social selves, pretend. Through the publication of a partial English translation of *Écrits* (1966) in 1977 (fully translated in 2006) he gained an English-language audience.

Lacan explored what can be said, or not said, in language as the medium of the unconscious, the unknowable other – the knowledge of which is yet, in his view, the foundation of all the human sciences. Meanwhile, however, his colleagues disqualified him as a training analyst in 1964, and Lacan founded the first of his own two schools, the École Freudienne de Paris. There was an extraordinary penetration of psychoanalytic ideas, which many French scholars had previously despised as attacks on reason, into French culture. In the 1970s, Lacan attempted, but failed, to contain a feminist response to his work, which accused him of mystifying the feminine as 'other'. Finally, in 1980, he unilaterally dissolved his own remaining school. All these splits, and the personal power Lacan exercised, emotionally burdened any statement in the French setting about analysis, language or the self.

Lacan's work reinforced the image among the intelligentsia of psychoanalysis as a discipline in the humanities and not a natural science, an interpretive and not an explanatory undertaking. Other French philosophers or analysts also brought out the richness of a hermeneutic approach to Freud. Paul Ricoeur referred to 'demystifying hermeneutics', which was taken up as an apt way to characterize disabuse of innocence in modern thought. Influenced by such writers, psychoanalytic discourse spread into the intense and often stylistically rebarbative world of literary, art and film criticism. Academic psychologists, however, generally simply ignored Lacan.

Psychoanalysis also contributed to political thought. Though Freud avoided active politics, he was not indifferent; he held scathing opinions about mass democracy, expecting a desire for a leader would emerge. In spite of this, however, in the 1920s his work entered into the rethinking of Marxism which followed the failed revolutions in central Europe after 1918. Max Horkheimer assembled a group of social theorists interested in psychoanalysis at the university of Frankfurt (hence 'the Frankfurt School'), which included Adorno and Marcuse and had links with socialist Berlin analysts like Siegfried Bernfeld and Fromm. These intellectuals developed views comparable with Reich's, though they expressed themselves in more sedate terms, using Freud's thought to link material conditions, social structure and individual psychological character. They argued that earlier Marxists had wrongly ignored the different psychological constitution of groups and hence the way in which political power, mediated by family and class, fabricates the individual's subjective world. Freud had described sublimation of libido as necessary for civilization; the 'Left Freudians', in contrast, argued that sublimation, whatever else it might be, is a political process which reproduces the existing structure of power in society as part of individual personality. Events on the world stage overtook these radical arguments in the 1930s, but they had their day in the 1960s, encapsulated in the slogan, 'the personal is political'.

Feminist critiques of Freud also related social structure to the individual psyche. Freud gave only belated and then never fully committed consideration to female as opposed to male sexuality and to the female child's position in the family. His work, in this regard, was conventional: male nature is the norm. He thought of femininity as grounded in the young girl's realization of her lack of a penis and consequent feelings of inferiority: 'The discovery that she is castrated is a turning-point in a girl's growth.' With maturity, he argued, the focus of a woman's sexuality moves to the vagina and, when she desires and gives birth to a child (especially a male child), a woman finds satisfaction through a substitute for what she has previously not had. Already in the 1920s, Horney had made a more positive evaluation of femininity. More radical critics followed the lead of Simone de Beauvoir who, in 1949, described how for male writers the man 'is the Subject, he is the Absolute – she is the Other', which is the essence of feminist anthropology. Critics turned to examine the social and psychological influences on attitudes to sexuality, including the attitude of Freud himself. It is therefore at first glance somewhat paradoxical that psychoanalysis offered special scope to women – their contributions were substantial. Freud attracted and enjoyed the company of brilliant women, such as Andreas-Salomé. His consulting room also made him unusually familiar with women's secrets; all the same, it has been remarked that his patients responded to him as a man and thus confirmed assumptions about sexual difference which he already held. When it came to generalizing about the social position of women, apart from the elite few, he was largely uninterested. Some later feminists, however, found in psychoanalysis the tools with which to understand the reproduction of social power along the lines of gender difference.

Freud's clinical studies produced a theory of childhood, and many teachers and child psychologists took analytic training or at least sought analytic insight. Psychoanalysis appeared to be a potent means to reveal the child's nature, to judge parenting and to grapple with the irrationality of disturbed children. Taking analysis to children required new techniques, since the talk in analytic sessions, on which everything else pivoted, had to be adapted. Anna Freud became an authority because of her work with children and her elaboration of the theory of defence, published as *Das Ich und die Abwehrmechanismen* (1936; *The Ego and the Mechanisms of Defence*), as well as because of her unique personal position. When she moved with her father to London, she continued her work and cared for child victims of the Blitz.

When the Freuds arrived in London, they were met by a small but thriving, if divided, analytic community, with Ernest Jones (1879–1958), a member of Freud's 'committee', the senior figure. As in Vienna, there were close connections with educational psychology and child welfare. Isaacs, for example, head of the department of child development at the Institute of Education, undertook analytic training. In addition, Klein, after leaving Berlin, further developed her independent theory of the child's nature, emphasizing the mother's position. She delved further into childhood than either Sigmund Freud or Anna Freud thought possible, using play to interpret how children relate to objects in their world. She thought that the child's attitude towards food makes it possible to understand the earliest months. The baby's relation to the mother's breast, she argued, is the prototype of all later relations. Violent love and hate experienced at this stage establish internalized objects and patterns which the child projects into every kind of relationship as she or he grows up. Object relations are thus the key to the emotional world. Then, during the war years, the followers of Anna Freud and of Klein engaged in an extended controversy, which centred on training, and this determined the future shape of analysis in Britain. There was an orthodox Freudian school, a Kleinian group and an independent group including Donald W. Winnicott (1896–1971) and W.R.D. Fairbairn (1889–1964). These last two groups became known as 'object-relations theorists', and their work spread, partly with the support of the Tavistock Clinic. In the Clinic, the work of Michael Balint (1896–1970) and John Bowlby (1907–1990) stressed the mother's role and reinforced links between child welfare and psychoanalysis. Analytic thought influenced people in management, social work and the caring professions. The focus on mothering and the roots of personality tied psychoanalysis into the postwar aspiration to build up welfare provision as the means to a fair and civilized society.

Freud and many of his followers were, in Nazi terms, Jewish, and it was easy for the Nazis to slander analysts' portrayal of the negative side in human nature as the preoccupation of a peculiarly Jewish science. There was a large-scale migration of German, Austrian and Hungarian analysts, especially to the United States, and they had a marked impact in fostering psychological society. It was not the case, however, that analysis disappeared in the Third Reich. A cousin of Field Marshall Göring, Matthias Heinrich Göring (1879–1945), headed the renamed German institute for psychological research and psychotherapy in Berlin, dubbed the 'Göring institute', and offered psychological treatment to the Nazi party elite. In this bizarre situation, analysts attempted to rephrase psychological concepts, formulated to deal with neurosis, as a language fit for self-styled heroes who, all the same, remained in Berlin during the war.

Psychoanalysis also had adherents in Russia. For two decades before the outbreak of the First World War, there was a lively community of psychiatrists in close touch with developments in Germany, Austria, Switzerland and France. Many of these doctors were highly critical of the tsarist system and linked the hardships of their patients to the failures of the state. They were generally open to the idea of using psychological techniques with patients, and in this context Nikolai Evgrafovich Osipov (1877–1934) and others introduced Freud's ideas. Osipov subsequently corresponded with and met Freud and Jung. He worked at the first outpatient service for neurotics at the university clinic in Moscow and in 1910, along with other young therapists, started the journal *Psikhoterapiia* (*Psychotherapy*), which took an eclectic view. These physicians served in the army during the war and were appalled by the absence of a co-ordinated medical service. Thus many of them welcomed the new Bolshevik state, hoping that it would organize an efficient, modern medical service, including provision for psychotherapy. Osipov himself, however, joined the emigration, and he was instrumental in spreading psychoanalysis in what had become Czechoslovakia. The leading Moscow psychoanalyst Moishe Wulff (1878–1971) also left, for Palestine. Psychoanalysis, however, continued to arouse interest, and indeed a psychoanalytic institute formed in Moscow was the first such institute to be state sponsored. The young A. R. Luria acted as secretary of the Moscow psychoanalytic society. In the early revolutionary years, psychoanalysis was one of a number of experiments in new ways of life, and, because it claimed to reveal true instinctual and material needs beneath bourgeois values, it could appear radical. Subsequently, however, psychoanalysis shared the fate of virtually all new forms of intellectual expression – elimination – with the declaration of the Great Break, the decision to industrialize and collectivize at whatever cost, in 1927. There was no formal decree, but it is easy to imagine the antagonism of Stalinists, bent on changing the world through collective action, to a bourgeois way of thought which indulged sexual sentiment and subjective problems.

The rapid popularization elsewhere of analytic ideas in the 1920s was part of a reaction against Victorian culture and hypocritical public denial of common private knowledge. In the words of a flapper in a play from 1927: 'Don't you even know, mother, that everybody's thoughts are obscenely vile? That's psychology.' Though Freud did his best to distance his movement from such farce, his own stoic pessimism did little to dampen down conviction that a beast lurks within the human breast.

This emphasis on what conservative opinion found 'vile' changed with the migration of analysts to the United States, especially through the work of Horney and Heinz Hartmann (1894–1970). With an American psychiatrist, Harry Stack Sullivan (1892–1949), they created 'ego psychology', a form of psychoanalysis which treated the ego as an original mental structure with its own positive powers. In tune with meliorist values, they argued that the psychic core of personality is a capacity to integrate innate drives and social pressures and thus to mature in a genuinely self-fulfilling way. Horney, reacting against Freud's gender bias, created a positive image of the nurturing qualities of the female ego. Sullivan, whose theoretical disagreements with her divided analysts and influenced the institutional growth of the field in North America, endeavoured to develop analytic work with psychotics. Freud had concluded that psychoanalysis was impotent before the psychoses, as these illnesses damage the capacity for meaningful talk. Sullivan, however, suggested ways to engage with even the highly disturbed psyche and to seek its integrative core. Hartmann argued, in *Ego Psychology and the Problem of Adaptation* (1939), for a conflict-free ego sphere, active in the ordinary functions of perception, learning and memory, which normally enables the self to adapt to the social environment. This work suggested ways to relate psychoanalysis and the experimental psychology of learning and adjustment, and there was for a while an attempt to create a unified theory. Hartmann wrote an influential paper on 'Psychoanalysis as a Scientific Theory' (1959), while the analyst David Rapaport (1911–1960) constructed a natural science model of the psyche which compared mental and physical energies.

Another émigré psychoanalyst from Europe, Erik Erikson (1902–1994), also stressed the existence of a conflict-free area of the personality. He studied the conditions which nurture or damage the child's emerging autonomy, which was an interest with personal roots as he had Danish parents, though he was the stepson of a German-Jewish paediatrician. He trained as a child analyst with Anna Freud and then went to the U.S. in the 1930s. His work on the life cycle, which, in *Childhood and Society* (1950), gained a large public audience, described a natural pattern of ego development. He focused on what he identified as the adolescent identity crisis, resolved in the normal course of things by adjustment to society. By the time Erikson published *Identity: Youth and Crisis* (1968), academic psychoanalytic and psychological approaches to development, along with psychotherapy, had grown close together. At the same time, analysts and analytic ideas influenced psychiatry in the United States, to an extent which happened nowhere else. Psychoanalysis thus merged into a culture of personal psychological growth, adjustment and fulfilment. By and large, optimism ruled: a normal pattern of development integrates all aspects of the psyche, including aggression, a trait which the Americans thought essential for competition but which they distanced from Freud's notion of the death instinct. Incompatible with the spirit of Freud's work but in tune with ideology, at times psychoanalysis and the search for personal growth degenerated into the fantasy of finding a personality independent of culture and economic conditions. This informed 'the me generation' of the 1970s and 1980s.

The spread of psychoanalytically informed culture in the United States was not without its critics. It was the butt of endless jokes and contempt from the populist heartland of the country, helped by the concentration of analysts in the cities of the East and West Coasts. There were also intellectual critics, notably Norman O. Brown (1913–2002), who in *Life Against Death: The Psychoanalytic Theory of History* (1957) attacked ego psychology for its disengagement from tragedy and the true personal autonomy which, in his view, stems from the death instinct. Herbert Marcuse reworked the argument connecting capitalist culture and personal repression, and his study of *Eros and Civilization* (1955) bore fruit in the 1960s. Radical critique turned against psychological and psychoanalytic descriptions of 'normal' development and 'maturity' as props of a politically sick society.

The psychological ideas and the movement which Freud had tried to shape became a sprawling, protean growth, which indeed had roots in much more than Freud's writing. It made an immeasurable contribution to psychological society. There was enormous middle-class prosperity in North America and Europe in the second half of the twentieth century, yet there continued to be a search to infuse that wealth with a soul.

## PSYCHOLOGY OF THE SOUL

Most of the early analysts were atheists. Nevertheless, hope that religion could recover its place as a form of spiritual healing led some Christians to turn to Freud, as they were also to turn to Jung, in order to understand psychic life. Behind this was a deep concern among Protestant pastors, before and after the First World War, that the churches had little to offer men and women in the modern world. One of Freud's most loyal supporters in Zurich was a pastor, Oskar Pfister (1873–1956), who believed psychoanalysis and religion shared the goal of sublimating instincts to higher ends. In the Netherlands, Britain and elsewhere in the 1920s, pastoral concern created an environment receptive to psychology – psychoanalysis included – and there was a turn to psychotherapeutic intervention under the rubric of 'healing', running together the business of healing the soul with the business of physicians.

Earlier histories of psychology left religion completely out of the picture. It appeared self-evident: as psychology grew *as a science* it left religious belief behind. This way of thinking, in my estimation, needs to be completely revised. For anyone interested in ancient and early modern sources of psychology or, as it may be, analogues to psychology, religious ways of life are central. Sensitivity to religion is also necessary for comparing Western psychology with other beliefs around the world about the soul, nature and the person. In addition, as a matter of fact, a field called 'psychology of religion' has existed since the last decade of the nineteenth century and, despite its vicissitudes, continues to exist. A concern with religious questions and spiritual experience has been a common motive for taking up, not opposing, new psychologies. There is a history of healing, including spiritualism, and of pastoral concern which merges, without a discernible boundary, into counselling and clinical psychology. Lastly, as I have argued discussing psychological society, public fascination with psychology has certainly not been limited to the pursuit of science in a narrow sense. Whether to label this concern as 'religious', the 'life of the spirit', the 'pursuit of humanity's well-being' or just plain 'sentimentality' is a matter for debate, but there can be no mistaking the widespread longing for a psychology which somehow addresses the human condition. The London Methodist minister Leslie Weatherhead had a large audience for books with titles like *Psychology and Life* (1934). A collection of Jung's reflections in English (in 1933) took the unabashed title, *Modern Man in Search of a Soul.*

Religion is a way of life and not just, or even not particularly, a system of belief. For many, what matters is religious experience and the quality of life which it makes possible. Just the same can be said about public appreciation of psychology: it concerns how to live rather than scientific explanation. Argument about the incompatibility of scientific and religious belief may therefore be somewhat beside the point. Ordinary people's interest in the quality of experience and the pastoral uptake of psychology both suggest this.

James delivered his Gifford Lectures in Edinburgh on 'the varieties of religious experience' (published 1902 and still in print). His approach was wonderfully broad and sympathetic, and his very sympathy with the subject-matter blurred the question, which many people, then and now, would ask: Are beliefs true? A number of more or less contemporary psychological studies turned to the empirical evidence of religious commitment, assuming that this would support rather than undermine belief. Some early psychologists, like Hall and Baldwin in the United States, and Hugh Crichton-Miller (1877–1959), the founder of the Tavistock Clinic, thought a healthy and natural development would lead a person to the Christian life. In early years, such was the Christian reputation of 'the Tavi' that it was known as 'the Parson's clinic'. James's *genevois* friend, Théodore Flournoy (1854–1920), like James interested in spiritualism, wrote about religion as a living human experience, open to investigation, like life generally, through psychological science. F.W.H. Myers published his studies which argued for contact between what he identified as a subliminal self and the spirit world. Among Catholics, as noted in connection with Cardinal Mercier, there was a self-conscious attempt to revive a Thomist psychology of the soul. Of course, some psychologists, in marked contrast, were anti-religious: Charcot in medicine, Freud and Jones among psychoanalysts, Watson the behaviourist.

Among Protestants, faith in religious feeling as the irreducible and authentic object of theological discourse built a psychological category into the religious life. This became significant in religious writing at the end of the nineteenth century, just when new psychologies were attracting attention. Liberal Protestants turned to feeling for the personal reality which would bring traditional faith back into contact with the realities of modern social and economic life. References to personality (in English) at first had spiritual connotations. A reconstructed picture of Christ pictured him as the model man, a man of true feeling, and an exemplar for ordinary people. There were even portraits of Christ as psychologist. It was this liberal theology, with its picture of Christ's personality, which the Swiss theologian Karl Barth profoundly criticized in a response to the First World War stressing the absolute transcendence of the Christian God, his otherness to humanistic values.

The supposed opposition of science and religion was a specific target of Jung's psychology, and this goes a long way to explain the recurrent appeal of his thought. Jung, from the beginning and entirely independent of Freud, had a large ambition to establish a *general science of psychology*. But by such a science he meant a consistent and true understanding of the full human being, which for him was the spiritually experiencing and motivated individual. He wrote: 'A psychology that satisfies the intellect alone can never be practical, for the totality of the psyche can never be grasped by intellect alone . . . the psyche seeks an expression that will embrace its total nature.' For Jung, also, psychology is *the* science, the foundation of knowledge of the human subject, on which all other knowledge finally rests. But what, for scientific psychology, is 'the totality of the psyche'? Jung endeavoured to answer this in an inward journey which it is easy to compare with myths of the quest for wisdom, the Grail or the Golden Fleece. In Jung's own terms, it was the quest for the archetype of the self which has the Christ-child as a symbol. It is not for me to say whether Jung was or was not a Christian – the argument is, I think, a religious one, and there are proponents of both views. But he did intend his science of the unconscious mind to speak to the religious sensibility.

Jung was born into a prominent and cultured Basel family, and he studied to become a doctor. One grandfather was a physician, the other a theologian who was said to have had visions, and his father was a Protestant pastor whom the son, at least, believed had religious doubts. While still a medical student, he worked with his younger cousin, Hélène Preiswerk, who was a spiritualist medium. In 1900, he gained an appointment at the Burghölzli hospital in Zurich, where he soon became clinical director and manager of the outpatient service, which involved hypnotic therapy, and he began to teach at the university. In 1906, he published the results of word association tests and this spread the notion of a psychological 'complex' or cluster of emotionally linked ideas. He also began to correspond with Freud after he was inspired, like his Zurich colleagues, by *The Interpretation of Dreams*.

Expansive letters and intense meetings reveal the stimulus Jung gained from Freud, to an extent a father figure, and the deep hopes Freud invested in Jung as an intellectual son and hero who would carry psychoanalysis to the international medical profession and wider world. Because of the stories told, which used to picture Jung as a follower of Freud, I need to emphasize that Jung, though the younger of the two men, came to the relationship with a strong sense of his own personal and professional worth, with a belief that he and Freud were pooling resources in an empirical project to understand the neuroses, and with his own way of thinking about the psyche. Freud's stress on the sexual libido was never fully acceptable to him, as he made plain to Freud, though Freud believed Jung would eventually have to accept the sexual theory because of the evidence. Jung used the language of 'will' or 'life force' to describe libido or psychic energy, in contrast to Freud's language which tended to describe the libido as analogous to a physical power. By 1911, even Freud had to face up to the differences of view, and in 1913 the rift became complete.

Freud believed he had discovered truths of world significance, and he perceived different views among those he thought his colleagues as defections which endangered his enterprise. Naturally, this caused deep resentment in those who pursued their own path. The relationship of Freud and Jung revealed much about the character of both men, as one can see that there always were different agendas beneath their apparently close friendship. Jung rightly claimed he had a different project and had never fully accepted Freud's views in the first place; his thought had autonomous roots in the nineteenth-century culture of psychotherapy, dream analysis, spiritual healing and the anthropology of myth and religion, as well as in psychology. He was an energetic and imposing man, a generation younger than Freud, a Swiss citizen and not Jewish. As Freud 'nearly said . . . it was only by [Jung's] appearance on the scene that psycho-analysis escaped the danger of becoming a Jewish national affair'. Trained as a psychiatrist, working at the internationally known Burghölzli, Jung was Freud's choice as the man who would take psychoanalysis to a wider medical audience. Freud therefore reacted angrily when Jung went his own way,

Jung had meanwhile begun to devote his clinical time to private practice. His work with psychotic patients in the hospital and with wealthy neurotic clients impressed him with the power and constancy of symbols produced by the mind as the expression of emotional pressures. He found the same vivid symbols in dreams and fantasies. He conceived of symbolism as representing the positive goals of the psyche – its intrinsic constructive as well as destructive nature. He thus opposed Freud's view of symbols as representations of the negative consequences of sexual repression. Whereas Freud thought in deterministic terms about the causal formation of unconscious content, Jung considered the psyche to consist of purposive forces known to consciousness in symbols and images. This became clear in his study, *Wandlungen und Symbole der Libido* (published first as articles, then as a book, 1911–12; translated as *The Symbols of Transformation*, though originally published in English as *The Psychology of the Unconscious*), written while still in touch with Freud. The book had all the excitement and disorder of a creative mind in full flow; in it, as the historian of the unconscious, Henri Ellenberger, wrote, 'Jung devoted more than four hundred pages to a mythological interpretation of a few daydreams and fantasies of a person he had never met.' He used reports published by Flournoy about a young American woman, known as 'Miss Frank Miller', who produced rich dream and fantasy images. In an interpretation of this material in which he drew on myth, philology and religious experience, Jung introduced a new language for dynamic unconscious powers. He did not deny the value of Freud's insights into repression but, all the same, he painted a picture of the unconscious with its own archaic, pre-individual structure. The result was the theory of *the collective unconscious*. Jung made this the basis for a systematic theory of personality in *Psychologische Typen* (1921; *Psychological Types*), where he introduced 'introvert' and 'extravert' to describe character types in which motivation derives primarily from inside or outside the psyche. The British psychologists Spearman and Eysenck subsequently took up these terms and they spread into the world of personality testing and popular psychology.

Jung also undertook a personal journey, a spiritual and psychological inquiry into his own psyche, which he likened to making shafts and tunnels into the depths of the mind. With his subsequent fame as a guide for the perplexed in soulless modern times, the journey was to give rise to much fantasy, mythology and fabrication. There is, however, a unique record, *The Red Book* (only published in 2009), of his fantasizing with mythology and symbolism, written in meticulous Gothic script and illustrated with symbolic paintings and illuminated letters as if it were a medieval manuscript. If in the eyes of some Jung emerged from his journey as a hero bearing spiritual insight, the record is of an extraordinary turn to myth and religious experience, transformed in a highly personal way, for truth about the mind.

In Jung's psychology, the foundation of human nature is inherited psychic energy, unconscious but continuously active in a purposeful manner proper to itself: there is 'a common psychic substrate of a suprapersonal nature which is present in every one of us'. This 'substrate' is the source of intuitions which express themselves as symbols. Since, for example, the wind as 'the breath of life' and as spirit is a symbol found in different cultures, this shows, Jung argued, that the unconscious psyche is collective. He studied recurrent patterns in the use of symbols, which he found in myths, religions, art and stories, and he discussed these patterns as *archetypes* or permanent structures of the collective unconscious. During the 1920s, Jung devoted considerable energy to non-Western cultures – for example, he visited the Pueblo Indians of New Mexico – and to Eastern religions, mythology and alchemy, in order to demonstrate the shared archetypal structure of the psyche. This included collaboration with the leading German sinologist and translator of the I Ching, or Chinese book of changes, Richard Wilhelm.

Jung, whose wife and later clients brought him wealth, resigned his university position at about the time he separated from Freud in order to give himself the freedom of private practice. His work with his patients, as with himself, pointed to a split between the conscious realm with its ethical motives, respectable emotions and rational knowledge, and the unconscious realm, which is amoral, asocial and non-rational. This split, Jung thought, emerges symbolically as pathological symptoms, in dreams and in glimpses of unachieved ideals beyond the surface of life. His approach to therapy followed from this. 'The pathological element [. . . lies] in the dissociation of consciousness that can no longer control the unconscious. In all cases of dissociation it is therefore necessary to integrate the unconscious into consciousness. This is a synthetic process which I have termed the "individuation process".' Belief in the possibility of integration was attractive in the confused years before and after the Second World War. Jung, however, also stressed that the unconscious energies are irrational and in many respects threatening and destructive; their successful integration can by no means be assumed. Jung, like Freud, thus conveyed a sense of imperilled civilization, and he too had an audience ready to believe in the tenuous hold of reason over the undiminished powers of the primitive psyche. Yet whereas Freud offered his readers the stoic injunction to develop scientific reason to lessen the intensity of suffering, Jung pictured the positive power of the collective unconscious to heal and transform.

It is striking how Jung's psychology commonly attracted people who were educated, professional and middle-aged – people outwardly successful but inwardly without direction. Jung understood this as confirmation of the existence in Western culture of a nearly catastrophic split between outward and inward life. The West, he wrote, preoccupied by knowledge and mastery of material things, lacks collective purpose and fails to give individual spiritual satisfaction. The culture as a whole is 'dissociated' from the collective unconscious. Jung moved on from being a young hero who journeys into the underworld to become a wise old man in a world eager for wisdom. Thus, in the 1930s, he commented on political events in the light of his psychology of the collective unconscious. As a German-speaking Swiss citizen, deeply imbued with German culture but not directly vulnerable to German politics, he was slow to appreciate the nature of Nazi power. It seemed to Jung that Hitler's appeal resulted from his genuine responsiveness to unfulfilled interests of the unconscious, interests unfulfilled because of the failure of post-Reformation Europe to achieve a collective symbolic life giving expression to spiritual needs. The result, Jung feared, was that dark and irresistible forces had overpowered the European mind.

Jung viewed what he wrote about events as objective contributions to the understanding of the modern psyche, that is, to psychology. To some later observers, however, his views were tantamount to support for National Socialism. Several factors may be borne in mind in considering this. Jung described inherited unconscious powers as having a character which reflects the particular history of a people or a 'race'. It therefore seemed natural and objective to him, as it did to many of his contemporaries – including some Zionists – to characterize the psyche in racial terms. He thought there was a Jewish psyche, the structure of which reflects the supposed rootless culture of the Jewish people. He assumed also that differences between Jewishness, Swissness, Germanness or whatever pointed to the need for different psychotherapy for different people. In the 1930s, whatever Jung's intention to be objective, all this was to play with political fire. Then, the April 1933 Nazification of German institutions resulted in Ernst Kretschmer's forced resignation as president of the German society for psychotherapy and the forced resignation of Jews from the society. Jung, the vice-president, responded by becoming president of a newly constituted international society for psychotherapy. His supporters argued that this put him in the best position to help those expelled and to preserve therapeutic ideals in the German-speaking world; his detractors argued it betokened his willingness to work alongside National Socialism and with its racial policy. Jung faced a moral dilemma. We may conclude, at the very least, that as a writer who had dismissed political activity as superficial in comparison to psychological understanding, he was ill-equipped to comprehend the political nature of his own role. He was also slow and reluctant to distance himself from the politics of racial stereotypes, since these stereotypes appeared to him to express an inner reality.

Both Freud and Jung (like Lacan later) claimed to have laid the basis for the objective science of the psyche. Freud continued ambivalently to draw on natural science for thought about the psychic sphere and opposed science to religion. Jung took religious experience to be central to psychology's subject-matter, and he presumed that whatever is psychologically true must also be religiously true, and vice versa. In relating psychology to culture, both made speculative use of belief in the inheritance of acquired characteristics – a commonplace belief in 1900 which became unacceptable during the 1920s, though more slowly in the German-speaking world than elsewhere. The new science of biological genetics simply did not allow for the idea of an acquired psychic inheritance, and this made both Freud's and Jung's approach to culture very vulnerable to criticism. Jung's work, indeed, acquired authority because of the insight and order he brought to clinical material, to comparative mythology and to seekers of an integrated and fulfilled subjective life, not as a contribution to natural science. For most of his psychotherapeutic followers and much of his public audience, the question whether his work was scientific was beside the point. Once again, then, we come up against the divergence in the twentieth century between academic and public expectations about what psychology is for. Beyond this, we come up against the recurrent tension between scientifically authenticated knowledge and the authority of personal or experiential meaning. Jung encouraged public belief in the significance of personal meanings, especially in his posthumous *Erinnerungen, Träume, Gedanken* (1962; *Memories, Dreams, Reflections*). In fact, though this text was published and read as autobiography, it was Jung's long-time personal secretary, Aniela Jaffé, who wrote much of the book and shaped it into an ideal story.

Running through this chapter has been an apparent paradox: rational, scientific investigation of the psyche demonstrates the limits of reason. The study of the unconscious supported psychologies open to the boundless range of human experience in religion, art, mythology and utterance of every kind, and not just to the circumscribed experience which appears amenable to the methods of objective natural science research. Taken to an extreme in the writing of some postmodern interpreters, inquiry into the unconscious has brought the Enlightenment to a close – though massive faith in the natural sciences demonstrates this simply is not so. Whatever critical judgment there may be about the credentials of the dynamic psychologists as scientists, there remains the historical and cultural fact that their writings have been the source of a rich language, free of stultifying mechanist conceptions, which people have taken up as a response to human desires. This language has not limited itself to matter or spirit, but, rather, held up the promise of a modern transcendence of the bifurcation – the language of the psychological.

## READING

If the extent of published work is an index, then psychoanalysis and psychotherapy, and their history, has a commanding place. Though published over 40 years ago, and of course in need of correction as so much detailed historical work has been undertaken subsequently, H. Ellenberger, *The Discovery of the Unconscious: The History and Evolution of Dynamic Psychiatry* (London, 1970), remains outstanding. More up to date is G. Makari, *Revolution in Mind: The Creation of Psychoanalysis* (London, 2008). For critical analysis of Freud's construction of his own history, M. Borch-Jacobsen and S. Shamdasani, *The Freud Files: An Inquiry into the History of Psychoanalysis* (Cambridge, 2011). Freud and virtually all the post-Freudians have had biographers and critical assessments. There is a highly readable study of Freud's relations with women: L. Appignanesi and J. Forrester, *Freud's Women* (London, 1993). For the tension in Freud's writing about reason, A. I. Tauber, *Freud, the Reluctant Philosopher* (Princeton, NJ, 2010), which re-asserts the power of Freud's thought in moral inquiry. For sexology, A. I. Davidson, *The Emergence of Sexuality: Historical Epistemology and the Formation of Concepts* (Cambridge, MA, 2001), chapters One to Three. On translation: D. G. Ornston, Jr, ed., *Translating Freud* (New Haven, CT, 1991); A. Phillips, 'After Strachey', *London Review of Books*, XXIX/4 (14 October 2007), pp. 36–8. For an overview of psychoanalysis in the U.S., N. G. Hale, Jr, *The Rise and Crisis of Psychoanalysis in the United States: Freud and the Americans, 1917–1995* (New York, 1995); for France, A. Ohayon, *L'impossible rencontre: Psychologie et psychanalyse en France, 1919–1969* (Paris, 1999); and for the Nazi period, G. Cocks, *Psychotherapy in the Third Reich: The Göring Institute* (New York, 1985). Lacan is famously impenetrable by what the French dub 'Anglo-Saxon' reason, but see M. Bowie, *Lacan* (London, 1991), and for a history written from a position within the events, E. Roudinesco, *Jacques Lacan & Co.: A History of Psychoanalysis in France, 1925–1985*, trans. J. Mehlman (London, 1990). There are alternative accounts of Russian psychotherapy in A. Etkind, *Eros of the Impossible: The History of Psychoanalysis in Russia*, trans. N. and M. Ribins (Boulder, CO, 1997), M. A. Miller, *Freud and the Bolsheviks: Psychoanalysis in Imperial Russia and the Soviet Union* (New Haven, CT, 1998), and I. Sirotkina, *Diagnosing Literary Genius: A Cultural History of Russian Psychiatry, 1880–1930* (Baltimore, ML, 2002).

Graham Richards has written with insight about psychology and religion, especially in 'Psychology and the Churches in Britain, 1919–39: Symptoms of Conversion', *History of the Human Sciences*, XIII/2 (2000), pp. 57–84. For the Dutch case, J. A. Belzen, 'The Introduction of the Psychology of Religion to the Netherlands: Ambivalent Reception, Epistemological Concerns, and Persistent Patterns', *Journal of the History of the Behavioral Sciences*, 37 (2001), pp. 45–68; and for the U.S., H. Vande Kemp, 'Historical Perspective: Religion and Clinical Psychology in America', in

*Religion and the Clinical Practice of Psychology*, ed. E. P. Shafranske (Washington, DC, 1996), pp. 71–112. Modern scholarship and Jung's search for a general psychology are detailed in S. Shamdasani, *Jung and the Making of Modern Psychology: The Dream of a Science* (Cambridge, 2003) and in his edition of C. G. Jung, *The Red Book: Liber Novus*, trans. M. Kyburz, J. Peck and S. Shamdasani (New York, 2009); also, 'Introduction: Encountering Hélène: Théodore Flournoy and the Genesis of Subliminal Psychology', in his edition of T. Flournoy, *From India to the Planet Mars: A Case of Multiple Personality with Imaginary Languages* (Princeton, NJ, 1994), pp. xi–li. For the sensitive matter of Jung's politics, see the papers by the Jungian analyst A. Samuels: 'National Psychology, National Socialism, and Analytical Psychology', *Journal of Analytical Psychology*, XXXVII (1992), pp. 3–28 and 127–48.

CHAPTER SEVEN

# Individuals and Societies

> The very being of man (both external and internal) is the *deepest communion. To be* means *to communicate.*
> Mikhail Bakhtin

## IDENTITY AND DIFFERENCE

'Most persons admit', Darwin observed, echoing Aristotle, 'that man is a social being.' Marx was more radical: human nature 'as it comes into being in human history – in the act of creation of human society – is the true nature of man'. There was a difference here. For Darwin, as indeed for modern biologists and psychologists generally, the reality is the organic individual who, as a result of early human evolution, has acquired fixed needs and capacities for a social life. For Marx, however, a person's nature is historical and social; being human is not to be human with sociability added but to be social through and through. Even the notion of being an individual is a social achievement.

Social psychology has most often been understood as the study of individual people in relation to each other. If people are social through and through, however, all psychology is in a sense social psychology, and indeed the field of psychology has no identity of its own strictly independent of social science, linguistics, anthropology and history. The usual position expresses a robust individualism – the individual is the political *and* scientific reality. The alternative expresses belief in collective life as a condition of there being a human individual.

The matter is not only abstract: there have been parallel psychological and sociological forms of social psychology. The psychological form took as its starting point the biological individual and treated social relations as an expression of individual capacities. The sociological form began with social structures and looked to account for individual action in terms of their consequences. For critics of the former, like the German social psychologist Carl F. Graumann (1923–2007), the field ended up with a de-socialized concept of the individual. For critics of the latter, the science failed to do justice to the individuality of the biological organism and its evolutionary formation. Psychological social psychology and sociological social psychology existed side by side in the U.S. in the twentieth century as two specialities with the same name but with different personnel, institutional support, methods and content. The psychological version took natural science as its model, and it became a laboratory science explaining outward social relations by reference to inner states like motivation, attitude and cognitive dissonance. The sociological version was more oriented towards collective entities like culture or class and explained actions in terms of the linguistic and symbolic life of historically constituted society. In Sweden, social psychology actually developed as a speciality within sociology not in psychology. In France, perhaps the most influential work on the individual's relation to society was done by mid-century historians, the Annales group, who envisaged a historical social psychology of shared *mentalité*.

The connection between social psychology and politics was an especially sensitive question in the Soviet Union. An account of this is not a usual feature of histories of psychology, but I think it belongs here because the Soviet state claimed a uniquely objective understanding of people's social nature. If anywhere, in principle, this is where scientific social psychology should have emerged. But it did not (though some think Vygotsky laid the basis for the science). Though I draw attention to the Soviet case, it is not because social psychology everywhere else was free from the political process. The psychological and social sciences expanded most rapidly in North America, Australasia and Western Europe, from the 1940s to the 1960s, in democratically minded political contexts. The human sciences and politics had a common agenda, as G. W. Allport, discussed earlier in connection with personality and the author of a standard history of social psychology, made clear in outlining the tasks of the psychological field: 'Whether social science, under proper ethical guidance, can eventually reduce or eliminate the cultural lag [between material and socio-political progress] may well be a question upon which the human destiny depends.'

At the beginning of the book, seeking the roots of psychologies, I described the Enlightenment ideal of understanding human nature in order to increase happiness and bring prosperity. Proponents did not differentiate psychology and social thought. Smith lectured on the sentiment of sympathy (which he did not refer to as a 'psychological' capacity), jurisprudence, language and 'the science of a statesman'. The very word 'association', which became so pervasive in descriptions of the ways sensations or ideas combine in individual minds, carried a reminder of collective realities. French writers distinguished a domain called *le moral*, encompassing social activity. In the German states, beginning in the late 1770s, a number of writers, including Herder, spread the word '*Kultur*' in order to describe the collective, social framework of education and custom which determines human character. In the same period, fascination with the origin and development of language – which, as a symbol system, is in its nature a social phenomenon – also took for granted the individual's social being. Nineteenth-century philosophical anthropology, including its materialist incarnation in Marx's work, transmuted all this into a historicized view of human nature. This understood being human in historical and not biological terms as a creation of the social life which language has made possible. It approached character and mentality as the achievement of time and place, not of universally distributed individual capacities of varying strength.

The obsession with human variety (individuals, classes, male and female, races, nations) built up a language of commonplaces about human relations. It was very widely, perhaps universally, assumed that character of mind differs; peoples – the Poles, the Irish, the Chinese, the Zulu – have a distinctive psychology. Such beliefs or, as it may be, prejudices surely still have a large place in public opinion, as jokes about the Englishman, the Irishman and the Scotsman attest. Enthusiasts for physiognomy, literary writers and portraitists had long combined pictures of individual character and the character of the group to which an individual belonged. During the course of the nineteenth century, both individual and group character was increasingly attributed to psychological nature. The nineteenth century was an age of rising nationalism and of empire-building, and there was much attention to what gives people an identity and excludes others. Imagination about racial psychology loomed large.

To cut a long story short, by 1850 or so, anthropologists, many of whom took pride in a material approach to human nature as a scientific replacement for nonsense about the soul, widely thought brain size and head shape correlate with psychological character. Paul Broca (1824–1880), for example, founded the Société d'Anthropologie de Paris in 1859 in order to support such a science. 'In general, the brain is larger in mature adults than in the elderly, in men than in women, in eminent men than in men of mediocre talent, in superior races than in inferior races . . . Other things being equal, there is a remarkable relationship between the development of intelligence and the volume of the brain.' Researchers made callipers to measure cranial dimensions, or they filled skulls with lead shot to determine weight as a marker of relative capacity. The Swedish anatomist Anders Retzius (1796–1860) devised the cranial index, the ratio between the side-to-side and front-to-back dimensions of the skull. He located original European 'broad-heads' among the Lapps and Basques; elsewhere, he claimed, more intelligent 'long-heads' had replaced them. In the extended struggle over slavery and over black and native emancipation in the United States and in Latin America, similar language played a large part.

Argument about the character of woman as a category, which became a hot topic with the women's movement for education, legal equality and, later, political emancipation, also brought in disputed facts about the brain and its nervous energy. A sometimes vituperative literature, in which physicians took the lead, stressed that the reproductive function makes women different, psychologically as well as physiologically, as they naturally use up in reproduction the energy which in men feeds mental life. In a similar vein, writers drew on the rhetoric of division of labour between head and hand to defend the naturalness of social classes. Conservative writers claimed the higher and lower classes, like the head and hand, have separate control and executive functions: 'refined' higher individuals make natural leaders, while 'coarse' lower masses provide labour. Languages of difference, at one and the same time physical and psychological, permeated every corner of society from the family to the empire. This built categories derived from social life into claims about what is objectively true in nature. It recreated social order as natural fact.

A colourful version of these beliefs emerged in Italy. Supporters of modernization, resident in the north and acutely conscious of what they perceived as the backward otherness of the south, turned to positivist natural science to combat Catholicism and ignorance. In this vein, the new professor of legal medicine and public hygiene in Turin in 1876, Cesare Lombroso (1835–1909), launched a determinist natural science programme to deal with social questions. He and those who thought like him devised complicated apparatus to measure differences in head size and physiognomy, collected information on what they believed were innate biological differences between people from the different parts of Italy and distinguished a physically distinct criminal class in men and a parallel prostitute class in women. Lombroso campaigned for legal science to become a branch of natural science, arguing that an expert scientific administration should deal with crime and prostitution as questions of pathology. During the 1890s, debates about degeneration, setting Lombroso's school, which stressed the existence of innate types, against French doctors and social scientists, who attributed more to environment leading to inherited disorder, filled the first international conferences on criminology. Lombroso himself went on to become, in 1906, professor of criminal anthropology in

Turin. Anthropology subsumed the study of people's psychological character. Studying different human types, researchers picked out a distinctive individual case (or number of cases) as the type-model and assumed that visible symptoms (in facial features, head shape and so on) reveal inner psychological essence.

The shift towards a physical anthropology of human psychological difference was never total, however, and there continued to be scholars, especially in Germany, who argued that the history of language was a more objective source of knowledge of the relations of different human groups than anatomy. Many intellectuals continued to think of historical culture, made possible by language, as the place to search for the causes and development of psychological nature. Thus in 1860, Hajim (or Heymann) Steinthal (1823–1899), with Moritz Lazarus, established a new journal, the *Zeitschrift für Völkerpsychologie und Sprachwissenschaft* (*Journal for Cultural Psychology and Linguistic Science*) in order to publish research about culture understood as the expressive activity of different peoples' psychological capacities. They looked to language and the collective activity language makes possible to learn about mind. The editors fostered the systematic study of human nature through psychology and the history of language and culture, believing this, not race, shows what sort of psychology a people has. Their subject was *Völkerpsychologie*, the psychology of people as social beings, the field which Wundt was to take up as his approach to higher mental processes. In a widely read book, *Philosophie der Unbewusstein* (1869; *Philosophy of the Unconscious*), Eduard von Hartmann discussed culture as the expression of unconscious ideas, revealing a common psychology, shared by all humankind, in spite of local differences. Studies on language and symbol systems thus went on in competition with physical anthropology. There was a considerable amount of work, often associated with nationalist aspirations, to record and classify the mental life of 'the people' or '*Volk*', visible, if rapidly disappearing, in traditional stories, customs, speech patterns, song and so forth. Many people were interested in local folklore, and this, combined with common language about national character, may well be one of the largely unexplored ways in which psychology – social psychology – became a public pursuit.

'Armchair' scholars depended for their notions of human diversity on travellers' reports. These travellers included both folklorists and anthropologists, like Adolf Bastian (1826–1905), who did as much as anyone in the 1860s to shape a coherent view of anthropology as a distinct science. Bastian spent more than twenty years travelling over much of the world, and though he travelled light and did not spend long periods in the detailed study of particular peoples, he contributed an enormous amount of information and inspiration to the disciplines of geography and ethnology in Germany. He helped to found the ethnological museum in Berlin (opened in 1886) and to organize the exploration, especially in Africa, which accompanied Germany's late participation in the scramble for empire. All through this, his aim was 'to find an adequate methodology for scientific psychology'. Ethnographic work, he argued, reveals the mental life of peoples, and the more primitive a people in an evolutionary sense, the more their culture reveals the basic universals of human mentality. Ethnography of primitive people will reveal the universals of human psychology; history will record the creation of difference. His argument, in tune with long-standing faith in the innate capacity for growth in the human mind, also recognized a cause ('*Geist*', spirit) for human purposes in the world. A generation later, Jung gave renewed prominence to this kind of psychology.

Bastian's beliefs illuminate a tension. His intellectual goal was to uncover the universals of human mentality; his practice, to a significant degree, was to provide tools in terms of which imperialists voiced belief in the contrast between primitive and civilized mentality. It was the science of evolution which, in theory, reconciled the positions, since the evolutionary outlook assigned people a common origin, and shared mentality, but also a distinct position at different stages of development, and different mentality. It was a matter for empirical science to determine what actually was universal and what was particular. It turned out to be far from easy to say. A century later, in the 1970s, cross-cultural psychology began to develop as a speciality, and its proponents believed research showed a range of psychological capacities (colour perception was the model case) to be species-wide. Bastian's generation, like Darwin's earlier, took this for granted while at the same time voicing the language of difference, of 'higher' and 'lower', in one utterance after another.

In all this talk in the late nineteenth century about human variation, the category of race united scientific and public language. Racial language claimed to discern psychological as well as moral and physical difference with a monumental degree of prejudice. It is perfectly possible to describe human differences without using the word 'race', yet a huge number of European and North American people, educated as well as uneducated, at times believed that race was not just the language of empirical science but the foundation in terms of which to understand human character in all its forms. This requires explanation. It is not enough to dismiss the authors who stressed natural inequality of races as simply guilty of misusing science – though they certainly did this too by the standards of later scientists (and hence Stephen Jay Gould published *The Mismeasure of Man*). The historical fact is that psychology in general acquired a substantial part of its subject-matter in public debate about difference between individuals and between groups, while social psychology as a particular field emerged in the course of debate about what biological and cultural conditions contribute to difference.

An illustration: Walter Bagehot, the editor of the *Economist* magazine in the 1860s, wrote a series of essays, which he called *Physics and Politics* (published together in 1872), about the hereditary English character. According to him, the English working man has acquired and now inherits a distinctively disciplined psychological nature. It was therefore safe to extend representative government (he wrote at the time of a Reform Bill) because deference to existing social structure has become bred into the English people. The argument implied – there was hardly need to spell it out – that the psychological character of peoples in other nations, and especially in British colonies, was not of the same quality.

Both theoretically and historically, the arguments over the disciplinary course of development for social psychology came to a head in connection with hopes and fears about the character of people in crowds.

## THE CROWD

The Paris Commune in 1871 violently disrupted order and property, and a bloodbath followed when the communards resisted the restoration of government. Contemporary images symbolized the loss of personal and social control: the horrified bourgeoisie imagined women from the commune, *les pétroleuses*, burning the city rather than surrendering. A decade later, Taine, in a celebrated history of 1789 and its aftermath, followed in this respect by Émile Zola's novel *Germinal* (1886), portrayed the crowd as an unleashed beast. In this picture, people in a crowd lose control and moral restraint; instead of making individual moral choices, they act automatically, imitating those around them. The crowd continued to be a visible presence in turbulent French politics – in strikes, May Day demonstrations, in the brief but meteoric rise to political prominence of General Boulanger and during the Dreyfus affair. There was an indeterminate fear in the vast cities of Paris and London, which expressed the trauma of lives undone by urbanization. There was agitation for democratic government, and in the eyes of democracy's many critics, this meant crowd agitation for government by the crowd.

As so-called mass society emerged, so did enthusiasm for a science which claimed to reduce the mass to order: a science of social relations. Taine did not just wring his hands in horror but argued the need for moral and political science. His friend and fellow historian, Émile Boutmy, commented: 'it was the University of Berlin that triumphed at Sadowa' (where the Prussians defeated the French in 1870, which was followed by the commune), and he went on to write that France's future depended on the education of its people, 'to give the people back their heads'. Taine introduced British empirical psychology to a French audience in a book on intelligence in 1870 and, with Boutmy, founded a school for political science which followed the example of Mill and Spencer in England and grounded politics on the science of human nature. Belief spread in the systematic study of action, both of the rational individual and the irrational crowd, as of utmost importance to modern politics.

As a result, crowd psychology flourished in the 1890s, especially in France and Italy, and it provoked a self-conscious attempt to formulate social psychology. In the same decade, however, Émile Durkheim (1858–1917) systematically set out to establish an alternative, sociological form of understanding of phenomena – crowds, shared beliefs and states of mind, as well as institutions (like religion or education) – which he understood to be social. From the start, he was committed to detaching social thought from the speculative evolutionary psychology of Spencer. For Durhkeim, collective life was a *social reality* in its own right, not to be explained as the sum of individual realities. The contrast and conflict between crowd psychology and Durkheim's sociology over the explanatory role of psychological as opposed to social categories posed the issues which subsequently divided psychological and sociological approaches in social psychology. If crowd psychologists ignored social structures and institutions, the Durkheimians retained assumptions about individual psychological nature in social theory (as Durkheim himself did in characterizing *anomie*, a state resulting from the lack of a secure framework of rules, which he used in his account of suicide and formless modern life). Indeed, Durkheim's followers, while contributing to social thought, gave considerable attention to 'the primitive mind' and the development of reason in culture. In the longer term, this was one starting point for the (sociological) social psychology of social representations advanced by Serge Moscovici (*b.* 1925) in the 1960s.

In Italy, there was also a literature on the crowd. Lombroso, Enrico Ferri (1856–1929) and others, centred on the Turin and Bologna law schools, studied the biological and material condition of the people, advocating legal reform on scientific lines. But whereas Lombroso identified a criminal type and accepted physical determinism, the crowd psychologists, like Ferri's student Scipio Sighele (1868–1913), argued that a person turns criminal as part of a crowd. It was even argued that the crowd itself may become a criminal entity. There was agricultural unrest and strikes by labourers, and the Italian lawyers (unlike the French theorists) cited the crowd as an extenuating circumstance for rioters when they appeared in court: the individual, as a member of a crowd, they argued, is less responsible. Some writers began to study groups as the positive expression of the people's collective life.

Crowd psychologists presupposed that the individual is the key to social order and disorder alike, since they thought individuals in groups are more likely to act in the way other people act than to make autonomous choices. They contrasted individual rational and moral choice with irrational collective conduct. Gustave Le Bon (1841–1931) and Gabriel Tarde (1843–1904) enriched this analysis and revealed what it owed to the debate about hypnosis between Charcot's Paris school and the Nancy school. Le Bon, in a hugely popular book, *Psychologie des foules* (1895; translated as *The Crowd*), and Tarde, more academically, in *Les Lois de l'imitation* (1890; *The Laws of Imitation*), highlighted 'suggestion', the power of external forces or other people to influence action, even without a person's awareness, and it became a key part of the imagery of the man or woman buffeted in the modern city.

Le Bon was egocentric, a misogynist and a rabid political elitist, yet one of the most important popularizers of science in France; though hardly original, his formulation of crowd psychology became the best known. For Le Bon, the crowd represents everything inferior: 'impulsiveness, irritability, incapacity to reason, the absence of judgment and of the critical spirit, the exaggeration of the sentiments . . . which are almost always observed in beings belonging to inferior forms of evolution – women, savages and children'. Men en masse, he argued, lose their capacity to chose, and so crowds are like women, constitutionally inferior and easily led. This explains the power of leaders and the fickleness of crowds, just as it explains the power of men over women. The outstanding leader, Le Bon argued, can sway the crowd. Contemporary politicians noted with interest that the crowds who excitedly called for General Boulanger to take power in the mid-1880s were unaware of the way a small group of wealthy tacticians had placed him in the limelight in the first place. Observers also believed that the radical May Day crowds which terrified conservatives at the beginning of the 1890s were, within a few years, baying against liberal Dreyfusards.

Whether the same proletarians formed the crowds on different occasions historians doubt, but contemporaries concluded that ignorant people followed the suggestion of the day. Le Bon did not doubt the astute leader could mobilize mass support. He turned an elitist fear of mass politics into an opportunity, pointing out that the superior man could hope to achieve control over the inferior mass. He had an audience: his work was translated into Romanian, Swedish, Turkish, Japanese and other languages, and he became a correspondent of the powerful. Le Bon thus helped shift the perception of the crowd as an uncontrollable rabble to a disciplined force. When he ran a salon in Paris at the beginning of the century – he was held in almost total disregard by the academic establishment – he gained an audience in the higher reaches of the French military, where officers were interested in sustaining order and effectiveness in a mass, conscripted army. After Mussolini led the Fascists to power in Italy in 1922, Le Bon sent him autographed copies of his works, which the leader claimed to have already read. Even before 1914, it is likely that Hitler, then a down-and-out painter in Vienna, absorbed Le Bon's message. Also in Vienna, Freud was another reader, and this played a part in Freud's thinking about aggression and mass psychology in and after the First World War.

Le Bon applied his analysis to any kind of social gathering, including parliaments, and not only to the Paris street crowd. The quality of reason declines, he thought, wherever men attempt to reach collective decisions, and he was virulently opposed to socialism as incompatible with rational decision making. It was left to Tarde to develop the arguments into a theory of social relations with more subtlety and greater liberality of purpose (though he too was conservative in politics). As head of the statistical department of the ministry of justice, he had a professional interest in criminology. Impressed by the evidence of hypnotic suggestion, he elevated the individual psychological capacity to imitate and thus share with others into a basic explanation of social life, not just of crime. In *L'Opinion et la foule* (1901; *Opinion and the Crowd*), he defined social psychology as the study of 'the mutual relations between minds, their unilateral and reciprocal influences', and he detached it from preoccupation with crowds and distanced it from vague notions of a collective mind. He suggested the value of research on 'the public'. Though Tarde discussed psychological factors, his goal was a social science, and he thus had to defend his position on two fronts. He argued in one direction against belief that social structure and social relations result from biological laws, opposing, for example, racial theories. He argued in the other direction against Durkheim, who rejected explanations of social solidarity by the psychological theory of imitation. The historical outcome, as a matter of fact, was for Durkheim's approach rather than Tarde's to acquire an institutional position in France, though Tarde did initiate work on social interaction and the formation of public opinion.

Baldwin brought together elements of German, French and British work in *Mental Development in the Child and the Race* (1894), in which he related theoretical research on mental evolution to the practical question of how children become members of society. He praised Tarde's laws of imitation, which he claimed to have understood independently, and discussed psychological growth as the process by which a biological individual becomes a social person. He thought there were three main stages: a passive stage in which children receive images; an active stage in which children assume the character of what they perceive – they imitate, Baldwin said, like 'veritable copying machine[s]' – and the social world selects or eliminates appropriate or inappropriate action; and a mature stage in which children comprehend the world separate from themselves. The self forms as an image of others, he argued, and the image of others forms as consciousness itself develops. This was one source for Piaget's and Vygotsky's later studies of development.

The same mixture of ideas was present in American social science at the beginning of the century. It was a professor of social science and a social reformer, Edward A. Ross (1866–1951), who spent most of his career at the University of Wisconsin, who published widely read books on *Social Control* (1901) and on *Social Psychology* (1908) which reflected current views on the crowd and on imitation. Not until the 1920s did social science and psychology separate in the U.S. along the lines which were to become entrenched for the rest of the century.

The popularity of Le Bon's *The Crowd* attests to the way people were attracted to explanations of social life in terms of individual psychological qualities and instincts. Belief in instincts, such as an instinct of imitation, played a large part in psychology until the 1920s, when it became clear to many that 'instinct' was as much in need of explanation as the psychological life for which it was the supposed explanation. When experimental social psychology developed in the 1920s, which U.S. psychologists consider the beginnings of the modern sub-discipline, it appeared a methodologically rigorous alternative to 'armchair' speculation about innate psychic structures. Also in these years, English-language social anthropologists reacted against speculative accounts of psychological stages in the evolution of society, and they structured the discipline around empirical field studies, in the manner of Malinowski's studies of the Trobriand Islanders.

In Britain, instinct featured in a distinctive psychological approach to social relations in the work of William McDougall (1871–1938). He became interested in psychology as a medical student, took part in the expedition to study 'the primitive mind' in the Torres Straits islands and then obtained a lone appointment in psychology at the University of Oxford, from where he moved to the United States (and provided support for parapsychology). He built his reputation on arguing that psychology of mind must have a physiological base, and he then published *An Introduction to Social Psychology* (1908), one of the best-selling English-language psychological texts of the interwar years. As an 'introduction', he described one after the other the innate dispositions which form character and hence account (in his view) for social relations. If his ambition was a philosophy of the place of humans in nature, an ethics and a social science, rather than a narrowly based psychology, what he conveyed was a picture of instincts determining experience and conduct. This, it appears, is what a large section of the public wanted and expected from psychologists, and McDougall, who was in fact an elitist in politics, became a spokesman for the commonplaces in terms of which ordinary people voiced their views on human differences. Describing inherited traits which were thought to be shared, McDougall and others referred to 'the group mind', 'a mental life which is not the mere sum of the mental lives of its units', and which, in its highest forms, becomes 'the national mind'.

It proved impossible to say clearly what an instinct was or to arrive at a definitive list of the instincts. In 1919 in a short paper, the U.S. psychologist Knight Dunlap (1875–1945) focused the criticisms. McDougall came to regret ever using the phrase 'the group mind' since it was so liable to misunderstanding. He never intended to say that there is a consciousness other than the consciousness of individual minds but only that there are similarities between individual minds, for which reason people respond in a similar way. Yet he and others described how this common structure creates a mentality which transcends the life of any one individual. This was the social reality which sociologists influenced by Durkheim attempted to understand in non-psychological terms, developing notions of *mentalité* and of collective representations.

In Britain, Graham Wallas (1858–1932), an active member of the left-wing Fabian Society and later Professor of Political Science at the London School of Economics, published *Human Nature in Politics* in the same year as McDougall's *Social Psychology*. His interest was in the contribution of social engineering to ordered political life. Pursuing this goal, more subtle in his understanding of politics than McDougall, Wallas included research on primitive and instinctual forces on the agenda of social science. A London surgeon, Wilfred Trotter (1872–1939), wrote on *Instincts of the Herd in Peace and War* (1916, expanding articles which first appeared in 1908–9), which presupposed a gregarious instinct as a positive force making it possible for people to live together, overcoming their selfish pursuit of pleasure. Psychology and social science, Trotter believed, should include comparative research on social animals – bees, sheep and wolves exemplified three types of socialized, protective and aggressive gregariousness. In contrast to McDougall and Freud, he invested his hopes for the future in the natural instincts of the human 'herd', though he too hoped reason would encourage leaders, people possessed of superior reason, to take charge. This kind of interest in relating social life and comparative psychology persisted into the 1930s, and Carl Murchison's *Handbook of Social Psychology* in 1935 had whole sections on 'infrahuman societies' and the 'social history' of the major human groups. Thereafter, however, experimental methodology turned social psychology into a more narrowly conceived field.

Freud, Jung, McDougall, Wallas and their contemporaries, albeit in different ways, assumed that there are inherited psychological structures acquired in the common evolutionary past. In the German-speaking world, it was common to envisage this as a kind of psychic, not material, inheritance, calling up a notion of the *Volksgeist* ('the spirit of the people') going back to Herder and the late eighteenth century, re-expressed in Wundt's science of the higher mental processes and in writings by others about the evolution of language, myth and religion. At the same time, however, German social theorists like Georg Simmel, Weber and Ferdinand Tönnies, though much interested in the psychological conditions of social action, turned to the more empirical study of social factors in the historical development of institutions. Simmel, for example, in a lecture, 'Die Grosstädt und das Geistesleben' (1903; 'The Metropolis and Mental Life'), which was to become a landmark of modernist perception, linked the conditions of life in the crowded and socially open cities to the breakdown of traditional communities and a feverish restlessness of individual experience and motivation. He made no attempt to keep social thought and psychology apart.

The separation, when it did come, was an Americanization of social psychology. In the United States, social psychologists began to identify themselves as specialists by giving people tests or putting them in the laboratory to see how conduct altered under observable conditions, and they extended this to include social relations. At the same time, however, sociology also expanded and claimed to serve the integration, or adjustment, of individual and society. The outcome was an unsatisfactory relationship between the disciplines of psychology and sociology.

## SOCIAL PYSCHOLOGY IN THE UNITED STATES

Both sociology and psychology acquired a modern disciplinary and professional form in the United States in the closing decade of the nineteenth century. There was no obvious reason why the study of relations between people should belong to psychology rather than sociology or vice versa. In the interwar years, however, approaches to human relations started to differ in significant ways in the two disciplines. For the most part, psychologists acted *as if* individuals have a non-social dimension, for example, in perception, memory or emotional expression, which gives psychology its core subject-matter. This led them to understand social psychology as the study of core activity in *special* social settings, for example, in friendship, school or work. Floyd H. Allport (1890–1978, the elder brother of G. W. Allport), in 1924, in what for many psychologists is the first textbook in the field, described social psychology as 'the science which studies the behavior of the individual in so far as his behavior stimulates other individuals, or is itself a reaction to their behavior; and which describes the consciousness of the individual in so far as it is a consciousness of social objects and social relations'. One might think, since there is no human behaviour or consciousness outside a social setting, that this renders social psychology the same thing as psychology in general. But it was accepted as a useful definition.

The absence of a predetermined boundary between psychology and sociology was evident especially at the University of Chicago where, following the philosophical leadership of Dewey and under the sociological leadership of Albion Small, there were high expectations about the practical benefits of studying people in society. It was the Chicago sociologist W. I. Thomas who, with Florian Znaniecki in their study of peasants who emigrated from rural Poland to Chicago, introduced the concept of '*attitude*' to describe personal orientation towards social conditions and values. Thomas and Znaniecki initially sought in universal human characteristics a psychological basis on which to erect a theory of social systems, and they turned to attitude as a linking concept. Znaniecki, however, like many other sociologists, subsequently decided that a social-psychological programme was not viable and began to use only social categories in his theoretical work. Sociologists turned to terms like 'role', terms defined by reference to social function not psychological content, to describe personal dimensions relevant to the examination of social systems. It was psychological social psychology which took over the term 'attitude'.

The position of social psychology 'between' psychology and sociology is vividly illustrated in the writings and legacy of George Herbert Mead (1863–1931). While surely the most profound American contributor to the field, psychologists ignored his work and sociologists took it up, but under the heading of 'social interactionism' (a term introduced by Herbert Blumer, who carried on Mead's lectures after the latter's death) and without the philosophy of which it was part. Mead also worked at the University of Chicago, and his social psychology can be understood as a working out of the implications of Dewey's 1896 paper on 'The Reflex Arc Concept in Psychology' in the light of German social philosophy of mind. Unfortunately, his early essays were in a forbidding, condensed style and the later books posthumous compilations from lectures. Mead argued that *all* psychological activity, even perception and emotion, has a social nature: it is not possible to talk about human experience or action (and he treated experience as a kind of action) independently of its social content. In a brief but incisive essay on 'Social Psychology as Counterpart to Physiological Psychology' (1909), Mead argued for a social psychology symmetrical with physiological psychology: mind is a function of society and not only of the physiological organism. He therefore attacked the then current opinion that imitation, conceived by analogy to a physiological reflex, is the means by which animals or children become part of a group. Instead, Mead argued, imitation presupposes knowledge of self and of others, and this knowledge itself emerges in a social process: 'the conduct of one form is a stimulus to another of a certain act, and . . . this act again becomes a stimulus at first to a certain reaction, and so on in ceaseless interaction'. Mead implied that we cannot assign meaning to mental states independently of the social whole within which, as he argued, they are social processes. In his view, psychological categories are social categories. For Mead, notably, there is no self to become socialized; the self itself forms in a social process. Most psychologists did not take this on board. Mead, it seemed to them, by thinking of the whole (society) as logically and materially prior to the part (the individual), had put the subject of social psychology outside the psychologist's domain.

An attempt to escape from dualism, whether of mind and body, centre and periphery in the organism, society and the psycho-physiology of the individual or the internal observer and the external world, underlay all Mead's work. His thought, like Dewey's, was fully evolutionary, concerned with persons in (social) environments as a process not as the interaction of entities. He also wanted to overcome the dualism of theory and practice: he attempted, for example, though without success, to conciliate in the large garment workers' strike in Chicago in 1910.

Knowledge of Mead's work diffused back into psychology through sociology in the 1950s and 1960s. He became known for the phrase 'the conversation of gestures', which denoted the social process of communication – not necessarily by language – that constructs the entities of self and other. The gesture (a physical or behavioural act, including speech) 'reacts upon the individual who makes it in the same fashion that it reacts upon another', and this develops consciousness and the individual as part of a social process. The continuous accommodation of individuals to each other, which begins in a young child as adjustment of movement, is, Mead argued, in maturity primarily a function of language and what he called 'significant symbols'. He described how we construct our psychological world through dialogue, in movement and language, with our surroundings at every moment of life. Not the laboratory but the everyday world is the setting in which we must study psychology. It was a programme for psychology which, to most experimentalists, appeared to have more in common with the observational work in the field of a sociologist like Irving Goffman than with natural science. Goffman's work, especially in *The Presentation of the Self in Everyday Life* (1959) and in his study of a mental asylum, where he had worked anonymously as a warder, attracted a large audience with its portrayal of individuals as actors – intrinsically social actors – in creating the world around them.

The North American alternative to Mead was exemplified in F. H. Allport's *Social Psychology* (1924), which fitted the subject into the psychology curriculum and suggested an experimental research programme on the behaviour of individuals in interaction with other individuals. Attitude and personality developed as organizing concepts, relating research to the public goal of individual social adjustment. The title of the leading U.S. journal in the field, the *Journal for Abnormal and Social Psychology* (the '*and Social*' was added in 1923) revealed social psychology's debt to the political context in which the occupation of psychology began to flourish, as the instrument with which to diagnose and correct 'abnormality' in individuals. The mental hygiene movement of the time encouraged studies on child development and personality, and many papers in the journal conveyed the clear message that personality is the key to social relations. G. W. Allport became an authority on the trait theory of personality and attitude, topics which set the agenda for much later work in psychological social psychology.

Belief in the integration of society through personal adjustment made it important to understand the formation of individual attitudes and opinions, for example, on race and labour disputes. The journalist Walter Lippmann used the expression 'public opinion' in 1922, and in 1936 George Gallup initiated a technique to measure it by random sampling (the Gallup poll). More importantly for psychologists, Thurstone, working at the University of Chicago, in 1928 introduced a technique to scale attitude, opening the field to the kind of quantitative analysis in which psychologists could claim special expertise. And, indeed, the pressure to be more precise and objective about physical variables, and hence to do experimental research on narrowly defined topics, increasingly set the agenda. For example, aggression research during the 1930s involved psychologists studying its threshold in children provoked to show it under controlled conditions in a laboratory. There was a notable shift from using race as an explanatory concept to studying opinion about race.

The demand of rigour, to give the science credibility, brought social psychology into the laboratory and led to the translation of human relations into behavioural variables. The psychologists, as a result, did not just simplify social relations in order to study them empirically but created a new kind of social relationship – the person in a laboratory. This was ironic: the public justification for social psychology was that it would provide information of use to businessmen, politicians, educators and anyone else interested in guiding people's behaviour in society not in the laboratory. Rigour and relevance worked in opposition. The Dutch psychologist Johan T. Barendregt later called this situation 'the neurotic paradox': a methodologically correct project is irrelevant to life; a project relevant to life is methodologically incorrect.

Many people were aware that disciplinary divisions were artificial and laboratory studies of limited value to industrial or social problems. Patrons who wanted results were frustrated. The largest response came from the Rockefeller Foundation, which made a multi-million dollar investment in the Yale Institute of Human Relations, founded in 1929 (and discussed earlier in connection with the behaviourist, Hull). The deans of the law and medical schools at Yale, along with Rockefeller, opposed what they saw as over-specialization, and they therefore established the institute as a large-scale experiment in integrating psychological, medical, sociological and anthropological expertise in order to unify social science. They sought a co-ordinated research strategy after the manner of large business corporations. As a result, a distinguished group of researchers, including the anthropologists Malinowski and Edward Sapir and the psychoanalyst

Erikson, worked there for a time. It is striking, however, that integration was very difficult to achieve, and to the extent to which it was achieved, it was as the formal science of psychology of Hull. Sociologists kept their distance from the institute, partly because of internal politics at Yale, partly because they sensed incompatibility between social and psychological levels of analysis.

The spread of psychology to business had begun before the First World War, but the link between academic studies and corporate interests consolidated after the experiments conducted between 1924 and 1933 in the Hawthorne plant of the Western Electric Company in Chicago. In the version of what happened promoted by psychologists, academic researchers set up experimental production facilities within the works and, surprisingly, found that physical changes in working conditions, like the lighting level, do not correlate with productivity. This led to the discovery that productivity is a function of the workers' attitude, and attitude itself appeared to emerge in the social process of establishing a group leader. As a result of the publicity given to the experiments by an academic at the Harvard Graduate School of Business Administration, Elton Mayo (1880–1949), who presented himself as the organizer of the studies, they became widely known and discussed. Mayo interpreted the results to show that productivity increases when workers get to know and like each other. Management, he argued, must look beyond engineering to personal relations. The experiment thus demonstrated the practical value of research in social psychology, and its outcome was thought to be a revolution in business practices, with the introduction of personnel management and worker-oriented redesign of work.

This version of events, save for the description of Mayo's influence on later business, does not stand up. The original experiments were part of a programme conducted under the auspices of the National Research Council and funded by electrical companies to set standards of high lighting levels in the interests of industrial efficiency. It was company engineers who started the experiments, and they were not at all blind to the effect of psychological factors; indeed, they set up the test situation precisely in order to eliminate psychological variables. Managers had for some time attempted to manipulate productivity with psychologically oriented schemes, such as payment by group rather than individual productivity. The fact that women workers engaged on repetitive tasks responded positively when given special attention and encouraged by observers expecting higher productivity was hardly a discovery of social psychology. Mayo, however, in his reports, transformed the test room into a type of laboratory, treated the workers' views as behavioural variables, took for granted the expert's superior understanding and assumed the rationality of management as opposed to workplace goals. These reports found an audience at a time when labour unions, responding to the economic depression, were militant, as they suggested that personnel management could lead workers to accept the rationality of management goals personified in a well-chosen leader. The Hawthorne experiments, therefore, became a model for the psychology of adjustment.

The Second World War overtook debates about social psychology both at the Yale Institute and in the business community. Psychologists and social scientists of all persuasions eagerly redesigned their research in order to contribute to victory. The war was a major stimulus to social psychology, spurring large-scale studies of both soldier and civilian attitude and morale and studies of attitude change. It also gave academics a taste for the organization of work on a corporate scale rather than as an individual pursuit. The volumes on the American soldier, published with S. A. Stouffer as the principal editor at the end of the 1940s, were a major fillip to social psychology. Indeed, the field expanded rapidly, as did the psychology profession generally, and psychological social psychology became an independent sub-discipline, later with separate experimental and applied journals and values pulling even the sub-discipline in different directions.

Before and during the war, there was a significant drawing together of interests in anthropology, social thought and psychology in the so-called 'culture and personality' school and in studies of 'the authoritarian personality', discussed in connection with psychological society. Such work reached something of an apogee in a study in which the British psychoanalysts Geoffrey Gorer and John Rickman, in *The People of Great Russia* (1949), correlated a supposed Russian predilection for a strong leader with experience of tight swaddling in early childhood.

Very influential for the long-term development of social psychology as a field was the work of Kurt Lewin (1890–1947), a gestalt psychologist who came as a refugee to the United States in 1933. Lewin was a member of the Berlin institute of psychology during the 1920s, and he attracted a distinctive group of students – women, Eastern European and Jewish – to study human relations. The experimental work of this group was reflective and innovative. The researchers accepted that the construction of an experiment is itself a social act which creates new psychological circumstances, and they therefore replaced the model of the scientist as the observer of nature with the model of the scientist as a social actor who studies other actors with whom she or he is in dynamic relation. Lewin's student from the Ukraine, Tamara Dembo (1902–1993), who also later became well known in the U.S., exemplified this approach in her studies of anger. She required her subjects to tackle a difficult problem while she observed their conduct and offered no help. She wrote up her results as a qualitative description of the social dynamics of anger, a form of writing not found in statistical studies of traits or attitudes. Lewin generalized such methods in order to understand human action through group dynamics.

Dembo's and Lewin's descriptions of actions somewhat resembled reports by phenomenologists of the meaning which actions have for a person's place in the world, and their descriptions went against accepted notions of objective reporting. Lewin, however, was not a phenomenologist; like the other gestalt psychologists, he believed that a scientist should go beyond description to establish general causes. He therefore analysed persons at work in groups in order to locate the most general features of the situation, and he related these general features in a formal theory of the psychological force field (using the metaphor of the physical field of forces). To this end, he used the mathematics known as topology, a mathematics suitable for the description of whole-part relations.

Lewin found it difficult to get a suitable permanent position in the U.S., and his methodological studies remained untranslated. Yet he had good research opportunities, high quality doctoral students and access to a network of friends and private funding agencies. From 1935 to 1944, he was a professor at the University of Iowa, where he developed experiments on group behaviour which began to attract the interest of social psychologists. With other socially concerned psychologists, he founded the Society for the Psychological Study of Social Issues in 1936. The war effort then redirected both the society's and Lewin's own activities. Already, in 1939, Lewin had put his study of group dynamics to practical use when his students worked on problems of productivity with the Harwood textile company in Virginia. This led to developing techniques for participatory decision making, bringing those who have to carry out decisions into the decision process and thus motivating them. This became part of what was hailed in the postwar years as a business revolution. It also encouraged 'action research', advancing knowledge by involving, not distancing, the researcher and the people studied. Both the manager and the scientist, Lewin thought, need to recognize the continuously changing dynamics of social relations, the context of individual action. At the same time, his paper, with R. Lippitt and R. White, 'Patterns of Aggressive Behavior in Experimentally Created Social Climates' (1939), provided a model of research which brought complex social phenomena into the laboratory.

During the war, Lewin became an advisor to the Office for Strategic Studies and the Office of Naval Research, and he was able to attract large funds. He gave thought to practical problems about individuals and social change, and this became part of a discussion about how to re-educate Nazis after their defeat. He elaborated on studies of the contrast between authoritarian and democratic group structures. Finally, the significance of Lewin's work received wider recognition when he moved from Iowa to the Massachusetts Institute of Technology (MIT) in 1944–5 to establish the Research Center for Group Dynamics. This centre had a major influence on social and industrial psychology. After Lewin's death in 1947, the MIT research team moved to the University of Michigan, where it became part of an Institute for Social Research, which was the largest centre for academic studies of organizational practice in the country.

Researchers studied techniques of leadership, motivation and conflict resolution within small groups, while Lewin himself established a technique for teaching group dynamics to future group leaders. In 1947, these techniques became the programme of the National Training Laboratory at Bethel, Maine, which trained leaders in how to produce organizational and attitudinal change. The training groups became known as 'T-groups', and they were the forerunners of the group techniques later familiar to people in many occupations – health care workers, for example. What went on in North America was closely followed in Europe. In London, the Tavistock Institute of Human Relations came into existence immediately after the war, and there Wilfred Bion (1897–1979) explored leaderless group dynamics from a psychoanalytic perspective. He also joined forces with professional bodies, especially social workers, to study and train people in group relations.

A pattern of ever expanding teaching and experimental research grew after 1945, which involved one conceptual model after another, generating enthusiasm but then fading as a new direction appeared. There was work, such as the influential studies of Leon Festinger (1919–1989) on cognitive dissonance, which emphasized that relations between people depend primarily on cognitive understanding. Other studies, some indebted to psychoanalysis, paid more attention to emotional drives. What gave the field an appearance of unity and direction was the experimental method, which treated individuals as independent actors and then studied their interaction in controlled circumstances and used statistical analysis to check the robustness of results.

One experiment, which stimulated enormous interest and debate, illustrates the pattern. In a paper on the 'Behavioral Study of Obedience'

(1963), Stanley Milgram (1933–1984) carried out an experiment in which he instructed college students to give electric shocks of increasing and finally of great severity to other participants who were out of sight but audible and whose ostensible task was to memorize words. The second group of participants were in fact accomplices of the experimenter and received no shocks. The experiment showed that students were willing to administer shocks when instructed by an authority figure, even while exhibiting distress at obediently causing pain. At first glance, Milgram appeared to have shown how easy it is to induce cruelty despite countervailing values and emotional distress. But several criticisms were levelled at the experiment. First, critics argued that it exploited the respect which ordinary people accord science and revealed the students' trust in institutions and experiments rather than anything about obedience and cruelty. Second, critics questioned the ethics of experimenting with human subjects. Milgram, in fact, in a process which he called 'debriefing', had talked after the experiments with his naive participants, informed them about the experiment and dealt with any emotional consequences. Nevertheless, ethical concerns subsequently entered more and more into the design and implementation of work in social psychology. This involved some recognition of the social dynamics of the experiment in a way which was foreign to many earlier behavioural studies. Third – a general criticism – what did experiments with U.S. students reveal about human motivation and the conduct of people around the world?

In the 1950s, there was a renewed attempt to unify the human and social science fields and to create a unified social psychology, this time under the banner of the behavioural sciences. It was not a success, and by the early 1970s a number of psychologists were referring to a crisis in social psychology. There was agreement about the key terms for the psychological analysis of social realities only within the narrow confines of particular branches of the profession. When psychologists tried to have influence in the wider world, they had to confront their own divergences of opinion, the different interests of different occupations and the messiness of political life. Moreover, largely independently of their efforts, psychological assumptions about 'economic man', an idealized representation of human nature, was embedded in public life and provided an intellectual rationale for the political and economic policies which dominated Western countries in the 1980s and then spread eastwards in Europe after the changes of 1989–91. Human nature, according to this political ideology, creates social relations through individual acts which consciously and rationally maximize individual material benefits. It was a way of thought which appeared to have far more clout than the studies of social psychologists.

There was precedent for a social and historical, rather than experimental, direction in social psychology in late nineteenth-century German historical sociology. This had a legacy, for example, in the work of the Romanian-born Norbert Elias, who published in the 1930s on 'the civilizing process'. Elias described the formation of early modern European character through the interaction of social customs, personal manners, bodily expressions and belief about the world. A pioneering study of the psychology of past people, Zevedei Barbu's *Problems of Historical Psychology* (1960), brought the principles of the Annales group of historians, encapsulated in Lucien Febvre's call for a history of sensibility, into contact with English-language interest in character and personality. The French historians envisaged studies of the long-enduring *mentalité* which comprehends time and geography, birth, childhood and death. This work, which took for granted Barbu's starting point that 'of all living creatures man alone is truly historical', pointed to historical knowledge as the subject-matter of psychology.

This was a long way from what most psychological social psychologists understood their subject to be. Their subject, by and large, continued in the 1980s and afterwards to be 'the individual' conceived independently of 'the social'. With the rise to prominence of the neurosciences, this individualist orientation was much reinforced. Meanwhile, however, in the Soviet Union (1917–91) large claims had been made for the achievement of an objective, truly social, science of man.

## SOVIET PSYCHOLOGY

The Marxist–Leninist aim for the Soviet state was to found it on the objective science of human action. The Soviet government claimed legitimacy as the political embodiment of the knowledge of the human world established by Marx. The contrast was with Western democracies in which the legitimacy of government rested on the expression of preferences by citizens, preferences not founded, in Soviet eyes, on objective understanding of human affairs but rather on the interests of class. In fact, Soviet apologists argued, Western states perpetuated non-progressive power structures not based on science. All this potentially made social psychology a very important field, but it also required Soviet social psychology to be of a piece with Marxism–Leninism.

There is a grim irony. Whatever the theory, for much of the time the model of the physical sciences dominated psychology, and this model substantially ignored people as social actors. The historical and political reality was that the close identification of the state with the Communist Party of the Soviet Union, which meant the Party exercising centralized power, resulted in a situation in which it was difficult, when not impossible, objectively to engage in the sciences which supposedly legitimated the state. It was least of all possible to carry out empirical social studies of human action. At its Stalinist extreme, the government tolerated no theory which the Party itself had not formulated: 'All efforts to think of any theory, of any scholarly discipline, as autonomous, as an independent discipline, objectively signify opposition to the Party's general line, opposition to the dictatorship of the proletariat.' The very argument which, in theory, made the Soviet system uniquely able to create a social psychology – its understanding of the materialist base of history – in practice prevented serious study of the subject. This kind of harsh mismatch between theory and practice, we may think, destroyed belief in the legitimacy of the Soviet exercise of power.

There is still a rich story to tell. As outlined earlier, psychology developed piecemeal before the Revolution, though Chelpanov's institute in Moscow, Bekhterev's psycho-neurological institute in St Petersburg and Pavlov's large team working on conditioning, based at the military medical academy, became well established. Several psychologists who later became important trained or worked with Chelpanov, notably K. N. Kornilov (1879–1957), Chelpanov's senior research assistant. Kornilov was a son of the rural intelligentsia, which in his case meant his father was a bookkeeper, and he projected himself to the Bolsheviks as a model of the hard-working student who raises himself and then returns, via educational psychology, to help his own class. If his work involved reaction-times and was of little service to the people, the rhetoric, conventional for the 1920s, reveals something about the social context in which people turned to psychology. Another researcher who worked with Chelpanov was P. P. Blonskii (1884–1941), an imprisoned student member of the Socialist Revolutionary Party during the uprisings of 1904–6 and then the first psychologist to support the Bolsheviks in 1917. He saw in revolution the opportunity to achieve pedagogic ideals and save people from their ignorance.

The collapse of tsarist power during Russia's war against Germany and Austro-Hungary, followed by revolution and civil war, was accompanied by huge material deprivation and, at times, anarchy in public life. Many people from the professional class, including natural scientists and doctors, left, but a certain number, including Chelpanov, Bekhterev and Pavlov, stayed on and sustained scientific work. In addition, some radicals in the early years thought that the Revolution had created entirely new social conditions, conditions which made possible a new human nature. A. K. Gastev (1882–1941) headed an institute for the scientific organization of labour in the 1920s, and he thought of the organization as the means to apply scientific knowledge of human beings as machines. Such radicals believed the materialist theory of human nature made it possible to re-engineer men and women and thus finally realize the dream of the 1860s generation and create 'the new man'. They thought that there need be no hypocrisy or frustration over the satisfaction of bodily needs and no intrinsic difficulty in each person learning to act for the public good in the light of objective knowledge. Blonskii threw himself into utopian plans to reconstruct not just academic culture but human nature itself, thus transforming the engineering ideal into social psychology. As a philosopher of education, he associated with Nadezhda Krupskaia, Lenin's wife and a force in the creation of a new educational system. With different goals in mind but in the same spirit of infinite possibility, the radical feminist Alexandra Kollontai advocated free love to liberate women from the economic dependency of marriage.

From 1923, radical plans for pedagogy and human engineering succumbed to political and economic contingencies. Any kind of practice was difficult, and there was more systematic attention to relations between the sciences and Marxism. Kornilov, for example, declared a mission to establish Marxist psychology at the 1923 congress of neuropsychologists. Already, beginning in 1918, the Socialist academy of social sciences had begun to train Party cadres in a Marxist understanding of human conditions. Renamed the Communist Academy in 1924, it was repeatedly in conflict with the academy of Sciences and pre-revolutionary scientific culture, especially when it tried to extend a Marxist evaluation of knowledge into the natural sciences, as it did, for example, in opposition to the new quantum mechanics. There were intense discussions among psychologists about the new character of science and about communist psychology in contrast to bourgeois psychology. The hidden side of this philosophical and ideological debate was a bitter struggle for scarce institutional and material resources, which caused researchers to fight for access to the Party – the sole power and dispenser of funds.

In the background were debates in Marxist philosophy which resulted, in the late 1920s, in A. M. Deborin's dialectical materialism replacing Bukharin's historical materialism (judged too determinist). All theoretical debate was then cut short in 1930–31 by the Communist Party subsuming theory and practice at every level in society to the transformation of the country known as 'the Great Break'. Many personnel active in this step had trained in Marxist practice at the Communist academy and had no time for the academic niceties of scientists trained before the Revolution. Out of these struggles, Pavlov's theory of 'higher nervous activity' emerged as the largest and most coherently organized programme for directing the course of scientific psychology.

The call to establish a Marxist psychology first came from Kornilov and from a physician involved in the mental health movement, A. B. Zalkind. Both were Party members and committed to the communist cause in a way which Bekhterev and Pavlov were not. In 1923, Kornilov wrestled the directorship of the Moscow institute from Chelpanov – as the Russian proverb puts it, 'when you chop wood, chips fly' – arguing that Chelpanov resisted psychology's reconstruction on Marxist lines and treated Marxism as one possible philosophy rather than the objective foundation for the sciences in general. Chelpanov could not shake off the accusation that he was a pre-revolutionary idealist. Kornilov headed the institute for the rest of the 1920s, but, in the early 1930s, in his turn, he was branded 'eclectic' and removed. His critics stated that he put together a mishmash under Marxist labels rather than an objective science able to carry forward the history then in the making under Stalin.

Whatever the ideological battles of the 1920s, one body of argument was, in the second half of the century, to achieve posthumous fame. This was the work of Lev Semyonovich Vygotsky (1896–1934), who had a significant impact on Western psychology from the 1960s to the 1990s as well as an influential place in the liberalization of psychology in the Soviet empire. Vygotsky was then called 'the Mozart' and the 'muffled deity' of Soviet psychology, which were phrases alluding to the brilliant versatility of a man who died young but whose reputation survived the darkest years of Stalinist brutality. As the separate Russian and English-language publication of his collected works in the 1980s and 1990s attests, he is a remarkable example of delayed influence.

Born into the family of a provincial bank official, Vygotsky had to overcome quotas on Jewish students to get an education. In Moscow, both before and after the Revolution, he was involved with a wonderfully inventive artistic avant-garde. His earliest work was as a literary critic, and in all his later work he attempted to accord a place to the expressive and aesthetic human consciousness. He probably welcomed the Revolution; certainly, within a few years he had thoroughly digested the Marxist classics. For reasons which remain obscure, in January 1924 Vygotsky appeared not in the guise of a psychologist of art but as a spokesman for a psychology taking consciousness as its subject. At the second psycho-neurological congress in Moscow, the young man delivered an electrifying talk which took apart (though he did not name names) the pretensions of both Bekhterev's and Pavlov's programmes to be the objective science of man, and he implicitly criticized the sympathy of Party ideologues with such work. He said: 'A human being is not at all a skin sack [a sausage] filled with reflexes, and the brain is not a hotel for a series of conditioned reflexes accidentally stopping in.' Instead, Vygotsky turned to Marxism as a philosophy potentially able to reconcile a psychology of consciousness with the materialist science of the body. Moscow psychologists welcomed this defence of a psychology of consciousness as it undermined the claims for attention from the rival Leningrad schools, and Kornilov invited Vygotsky to join the institute.

Both the talk and the invitation may have come at the instigation of two young researchers, A. N. Leont'ev (1903–1979) and A. R. Luria (1902–1977), who were major figures in the revival of non-Pavlovian psychology and neurology in the 1950s. Leont'ev became known for his studies of child development, carried on through the 1930s, and Luria for his research on brain damage (a speciality called 'defectology' in the Soviet Union) on which he worked extensively during and after the war of 1941–5. Leont'ev, like Luria evacuated to the Urals, also worked with medical cases during the war.

Luria's reminiscences of the mid-1920s give a vivid idea of the turbulent social and intellectual setting: 'Instead of cautiously groping for a foothold in life, we were suddenly faced with many opportunities for action . . . An entire society was liberated to turn its creative powers to constructing a new kind of life for everyone. The general excitement, which stimulated incredible levels of activity, was not at all conducive, however, to systematic, highly organized scientific inquiry.' In the midst of this, Vygotsky took up dialectical philosophy in order to integrate a theory of biological development and of historical consciousness. In a short number of hectic years, years increasingly disrupted by tuberculosis, he worked in pedagogy and clinical psychology, he designed a psychological field study for rural Uzbeks, which involved philosophical questions about methods, and he did research on developmental psychology and in defectology. He wrote a large-scale theoretical text, *Istoricheskii smysl psikhologicheskogo krizisa* (*The Historical Meaning of the Psychological Crisis*) in 1926–7, which included a cosmopolitan review of competing philosophies but which remained unpublished until 1982. His studies of developmental psychology culminated in a series of essays, edited into what became his best-known work, *Myschlenie i rech'* (1934; *Thought and Language*, translated into English in 1962, again in 1986 and in 1987).

Vygotsky divided the child's early development into a pre-linguistic stage, for which biological knowledge is necessary and in which the child possesses only sensory and emotional consciousness, and a linguistic stage, in which the child interacts with historical culture and thereby acquires both language and the capacity for thought. He argued with the developmental theories of Piaget, and in this specific context his ideas first became known in the West, marginally in the 1930s and then substantially in the 1960s. This knowledge of Vygotsky as a developmental psychologist was at the expense of his broader social psychology, which concerned the interaction, mediated by language formation, between the individual as a biological organism and historically specific culture. Vygotsky, however, was unable to do much systematic socio-psychological research. He dreamed of a Marxist social psychology as the basis of a unified psychology, integrating biological and linguistic processes, but almost none of this appeared in *Thought and Language*. Moreover, the initial abridged English-language translation made Vygotsky's work appear simply a contribution to Western debates on developmental psychology. Though his work was cut short by his death, it was still the most serious attempt during the interwar years to achieve a Marxist psychology, even if this was not perceived or accepted in the Soviet Union in the 1930s and in the U.S. in the 1960s.

Forces which went far beyond misrepresentation came to dominate scientific life in the USSR. For several years before his death, Vygotsky, like other psychologists, faced criticism of his work for not adequately embodying goals defined by the Party. Any intellectual or artistic undertaking was subject to such criticism in the 1930s: the very act of standing outside the Party, even for purposes of debate or aesthetic expression, was taken to be opposition to the Party and its claim to have an objective hold on the human condition. More prosaically, Party activists, faced as they were by the demands of the five-year plans for accelerated industrialization, were impatient with theory of any kind and disdainful of highbrow science. These Party officials tolerated theory less and less, and they questioned the practical expertise of psychologists and even the value of psychology as a distinct discipline.

Pavlov's school, with its well-supported infrastructure and description as 'physiology', a subject which was politically more neutral than psychology, stood out as a relatively thriving area of research and was able to maintain continuity in its work. Most other areas, like psychoanalysis, suffered. One area which did expand for a while was called 'pedology', a term which covered research and practice on child development and the use of tests in schools. In the mid-1930s, pedologists argued that tests were valuable because they made it possible to pick out intelligent future leaders from the uneducated children of the people. Teachers resented this intervention as it detracted from their own judgment about children. Moreover, in practice, the tests identified children with incapacities, and this led to an alarming rise in the number of children described as 'defectives'. As in every other country, the practical consequence of testing was to favour children from the more privileged social strata. Argument about these questions led, in 1936, to the government shutting down pedology, and this adversely affected the prospects of many psychologists.

Party activists did not think the Soviet 'new man' needed psychology. In the political circumstances of the time, it was all too easy to equate the ideal type with the loyal supporter of Stalin and the Party. If it was possible, by the exercise of will and consciousness of historically constituted human nature, to leap beyond existing material circumstances, there was no place for psychological or social science. Historians working in the archives have also uncovered evidence about the struggles for resources and positions beneath the vague but coercive public language. Some things are also known about individual lives. Vygotsky died in 1934, aged 37, mentally distressed at the marginalization of his life's work. His colleague Luria moved sideways from psychology, went to medical school and worked in the politically less contentious area of neurology. In 1933, Leont'ev relocated with others to somewhat politically less conspicuous positions in Khar'kov, where, after Vygotsky's death, he continued to work on a Marxist theory of child development. At this time, curiously, he also worked for his higher degree (habilitation) on experiments which appeared to show the acquisition of the capacity to perceive colour through the hands.

Even as psychological activity looked threatened, a new philosophical voice came to the fore. How S. L. Rubinshtein (1889–1960) achieved prominence is not clear, but he became the spokesman for a politically acceptable dialectical psychology linking the biologically evolved body to historically situated consciousness. He was a professor in Odessa before his 1932 promotion to head the psychology department in the Herzen pedagogical institute in Leningrad, from where he published 'Problemy psikhologii v trudakh Karla Marksa' (1934; 'Problems of Psychology in the Works of Karl Marx') as well as a textbook which described psychology in appropriate Marxist language and kept alive teaching in the subject. Interestingly, Rubinshtein, if in somewhat vague terms, turned to the recently published manuscripts of the young Marx for his basic concepts. As he wrote, 'the point of departure for the reconstruction [of psychology] is the Marxist concept of human activity', and he then went on to quote

Marx's view that Hegel had revealed 'objective man, true, actual man, as the result of his own labor'.

Western Marxists later grappled with the same issues, and this had an influence in the psychological and social sciences. Marx himself had drawn on political economy and eighteenth-century theories of the stages of human progress, largely ignoring psychology. Marx's followers were interested in what people do, the world of economic and political relations which people create, not the subjective world. Yet, as readers of Marx's writings as a young man were to stress, when these writings were published and became known, beginning in the 1920s, the heart of Marx's criticism of capitalism was that it denied the development of individual human potential. At the back of Marx's thought remained the dream of *Bildung*, the development of the mental world and its expression in culture as the true human goal. In practice, he focused his energies on attacking the political conditions which, he thought, made that goal impossible, rather than developing a theory of human nature with a place for psychology. Many later writers, however, equally committed to social progress, were to think Marx erred in ignoring psychology. We need, they argued, a science, a social psychology, which studies how individual mental life and social structures are elements in a common process.

Rubinshtein won a Stalin Prize, was promoted and founded a chair of psychology in the philosophy faculty of Moscow University in 1942. Between 1946 and 1949, however, virtually all forms of psychology, including Rubinshtein's, faced criticism. The only exception stood in name on Pavlov's work. The Party then dogmatically confirmed support for Pavlov's science as the natural science of man in three all-Union congresses, in physiology (1950), psychiatry (1951) and psychology (1952). These congresses were characteristic of the final paranoid years of Stalin's rule (he died in 1953). In the aftermath of the war to end the German invasion, a new wave of terror and anti-Semitism gripped the USSR, and show trials and prison camps paralysed the Soviet dominated half of Europe. Isolation, suffering and the Cold War fostered an extreme chauvinism about Russian science, and public speakers stressed Russia's unique historical contributions to scientific progress. Psychologists constructed a genealogy to link the nineteenth-century physiologist Sechenov to Pavlov, and Pavlov to a dialectical science of being and consciousness, and they elevated this into a dogma which had a lasting influence on Soviet writing about the history of psychology. Psychologists feared that the power given to the heirs of Pavlov's physiological research threatened to eliminate psychology as an independent subject. However, they resisted this extreme consequence.

Within a few years of Stalin's death, psychologists were able to broaden the research agenda, though it was still necessary to make appropriate obeisance to Pavlov in public. Rubinshtein returned to prominence in the second half of the 1950s with a dialectical study of *Bytie i soznanie* (1957; *Being and Consciousness*). B. M. Teplov (1896–1965) elaborated part of Pavlov's thought into a systematic typology of human character, but he also published a textbook which clearly defended the autonomy of psychological from physiological topics. By 1960, Soviet scientists had re-established regular contacts with the West, and in the area of brain research this was with a real sense of discovery and excitement on both sides. Many researchers later commented on the stultifying effects of the decisions in support of Pavlov's science. Most damage was probably done in central Europe, most of all in East Germany (the DDR), where ideological controls were tight and where scientists, distant from the elite centres of Moscow and Leningrad, were unable to decide what if any leeway there was in the research agenda. Nevertheless, psychologists found ways and means of doing much non-Pavlovian work; the central European capacity for constructive survival under repression became legendary.

Leont'ev, Luria's and Vygotsky's colleague in the 1920s, had particular influence on a new generation of psychologists in Russia. He returned to Moscow in the late '30s, and, in the late 1940s, began to be the dominant influence on Moscow psychology. It was the institutions of pedagogy (and hence the Academy of Pedagogical Sciences), rather than those of natural science (and hence the Academy of Sciences), which provided a home for the expansion of psychology. To this day, no psychologist has been a full member of the prestigious Academy of Sciences. Nevertheless, in 1966 Leont'ev established the first separate faculty of psychology in the Moscow State University, and thereafter, till his death in 1979, he was its dean. He thus trained the generation of psychologists who acquired prominent positions as the Soviet Union entered the period of changes in the 1980s. Though many of these psychologists reacted against Marxism, Leont'ev himself had begun to publish a Marxist theory of development in the 1950s and, for example, to experiment on memory in order to show how the capacity of memory changes with the social activity of which it is part. He called his work 'activity theory', aiming for a unified account of action, the real activity of people in the historical and social world, and thus escaping from the explanatory terms 'mind' and 'body' of an untenable dualism. What people do is not a function of mind or body but of a process in which they engage with surrounding conditions. This proved to be a constructive basis on which to support both experimental work on development and satisfy the requirement of *praxis*, in order that science be a tool with which to change the world.

Many scholars contributed to the liberalization of academic and professional life which, with perestroika (reconstruction), grew into a flood after 1985. For example, the literary theorist Mikhail Bakhtin had a delayed but profound impact, in the West as well as in the Soviet Union, on interdisciplinary thought linking psychology to studies of language, reason, history and culture. Among psychologists, Mikhail G. Yaroshevskii (1915–2002) was responsible both for the first post-perestroika edition of Freud and for beginning more open-minded approaches to the history of Russian psychology.

Marxism–Leninism in the Soviet Union purportedly made objectivity in the psychological and social sciences possible, and yet the exercise of power destroyed objectivity and even, at times, threatened to eliminate the sciences themselves. Much of Soviet science singularly failed to research relations between individuals and society because the relation was, in fact, a matter for centralized political decision not science. (This was hardly unique to Russia: Western policies, for example, on imprisonment, frequently reflected political rather than social science judgments.) Thus the country did not equip itself well in scientific terms to face its own problems. The consequences were evident also in the confusions of the post-Soviet period, in a too-ready belief that capitalist economics would solve social problems and Western psychology heal personal lives. Yet the work of both Vygotsky and Leont'ev suggested Marxist alternatives. Both in Russia and in the West, there were psychologists, especially followers of Vygotsky, who looked here for a truly social psychology, a psychology able to show how society, as a historical process, is constitutive of psychological capacities and action.

People critical of the Soviet system might be tempted to explain its impact on the psychological and social sciences, especially its substitution of slogans or abstract theory for research, as the aberration of a totalitarian regime. But the attempt to create a unified science of the individual in society has everywhere been part of the political process. For many researchers in social psychology in the United States and elsewhere, it seemed natural to take the biologically independent individual as the starting point. The political and economic individual taken for granted in the public sphere reappeared as the 'natural' experimental subject. In practice, experimental social psychology examined individuals, usually students, in the laboratory setting, and recorded results as statistically significant variables. Psychologists rarely analysed this work itself as historically situated action or studied the social and historical nature of their experimental subjects. They conceived of practice in terms of the adjustment of the individual not of social change. Mead, who questioned the intelligibility of describing individuals independently of the social world, phenomenologists who rejected the value of quantitative experimentation governed by statistical models, and radical psychologists who believed psychological well-being required social change, all remained marginal to psychological social psychology.

All the same, there is something especially sobering about the Soviet experience. Revolutionary idealists set out with Marx's vision of human enlightenment to fulfil human potential through objective knowledge of people as part of the material world. They failed, and this failure produced untold suffering. Was this an indictment of the Enlightenment dream, surely a glorious dream, to end misery through knowledge? Scientists think not. Yet the historical fact remains that the most sustained experiment in constructing a new world, with the stated aim of founding policy on the science of human beings, so singularly did not succeed.

## READING

A readable introduction, highlighting the eighteenth-century roots, is G. Jahoda, *A History of Social Psychology: From the Eighteenth-Century Enlightenment to the Second World War* (New York, 2007); also *Crossroads between Culture and Mind: Continuities and Change in Theories of Human Nature* (New York, 1992). Essays by R. M. Farr, *The Roots of Modern Social Psychology* (Oxford, 1996), raise questions about the shape of history of the field; also C. F. Graumann, 'The Individualization of the Social and the Desocialization of the Individual: Floyd H. Allport's Contribution to Social Psychology', in *Changing Conceptions of Crowd Mind and Behavior*, ed. C. F. Graumann and S. Moscovici (New York, 1986), pp. 97–116. M. C. Carhart, *The Science of Culture in Enlightenment Germany* (Cambridge, MA, 2007), lays out a new historical basis for the concept of culture.

On Italy and Lombroso, D. Pick, *Faces of Degeneration: A European Disorder, c. 1848–c. 1918* (Cambridge, 1989); also, P. Becker and R. F. Wetzell, eds, *Criminals and Their Scientists: The History of Criminology in International Perspective* (Cambridge and Washington, DC, 2006). On Bastian, K.-P. Koepping, *Adolf Bastian and the Psychic Unity of Mankind: The Foundations of Anthropology in Nineteenth Century Germany* (St Lucia, Queensland, 1983). I refer to S. J. Gould, *The Mismeasure of Man* (Harmondsworth, 1984) for its demolition of the scientific pretensions of late nineteenth-century physical typologies. There is a rich literature on the crowd, including S. Barrows, *Distorting Mirrors: Visions of the Crowd in Late Nineteenth Century France* (New Haven, CT, 1981), J. van Ginneken, *Crowds, Psychology and Politics, 1871–1899* (Cambridge, 1992), R. A. Nye, *The Origins of Crowd Psychology: Gustave Le Bon and the Crisis of Mass Democracy in the Third Republic* (Beverly Hills, CA, 1975), and D. Pick, 'Freud's *Group Psychology* and the History of the Crowd', *History Workshop Journal*, 40 (1995), pp. 39–61.

For the United States, there are standard descriptive accounts in G. Collier, H. L. Minton and G. Reynolds, *Currents of Thought in American Social Psychology* (New York, 1991), and E. R. Hilgard, *Psychology in America: An Historical Survey* (San Diego, 1987). Kurt Danziger's work (see reading for chapter Four) has turned social psychology to reflect historically on itself. On Mead, including the German influences, H. Joas, *G. H. Mead: A Contemporary Re-Examination of His Thought* (Cambridge, MA, 1985). For social psychology's social engagement: J. G. Morawski, 'Organizing Knowledge and Behavior at Yale's Institute of Human Relations', *Isis*, LXVII (1986), pp. 219–42; R. Gillespie, *Manufacturing Knowledge: A History of the Hawthorne Experiments* (Cambridge, 1991); J. H. Capshew, *Psychologists on the March: Science, Practice, and Professional Identity in America, 1929–1969* (Cambridge, 1999); G. Richards, *'Race', Racism and Psychology: Towards a Reflexive History* (London, 1997).

I. Sirotkina and R. Smith summarize the history of Russian psychology in 'Russian Federation', in *The Oxford Handbook of the History of Psychology*, ed. D. B. Baker (New York, 2012), pp. 162–91, drawing on Sirotkina, 'When Did "Scientific Psychology" Begin in Russia?', in *Physis: Revista internazionale di storia della scienza*, XLIII (2006), pp. 239–71, and on Joravsky (see reading for chapter Two). Amidst the material on Vygotsky, R. van der Veer and J. Valsiner, *Understanding Vygotsky: A Quest for Synthesis* (Oxford, 1991) is both historically informed and critical. For the idealism of the Soviet project, R. Bauer, *The New Man in Soviet Psychology* (Cambridge, MA, 1952), retains interest.

CHAPTER EIGHT

# Where is It All Going?

> The acquisition of knowledge of mind and the mental [is] a component part of our more general socialization, involving as it does being . . . inducted into both our society's culture as well as into 'the common behavior of mankind'.
>
> Jeff Coulter, using a phrase from Wittgenstein

## PSYCHOLOGY AFTER 1945

There was a leap in the sheer quantity of psychological activity, so much so that one might almost call psychology a science of the second half of the twentieth century. There were increased numbers of psychologists in academic research and even more in clinical, educational and occupational psychology; and much of public life acquired the characteristics of psychological society. Even were history to restrict itself to 'the professionals', it would still be necessary to deal with vast numbers of people. Between 1920 and 1974, 32,855 doctorates were awarded for psychology in the U.S., 5,000 more than in physics. At the beginning of the present century there were over 150,000 people affiliated with the APA, while in the Netherlands there was perhaps an even greater number of psychologists per head of population. In some years, psychology was the most popular of all student subjects. A glance in a bookshop shows what hopes the field continued to foster, almost as if Cattell's brazen prediction in 1920, 'it is . . . for psychology to determine what does in fact benefit the human race', had become true.

Not just the scale but the variety is striking. In recent book catalogues, I find: *Rationality and the Pursuit of Happiness*; *You Were always Mom's Favorite*; *Psychodynamics for Consultants and Managers*; *Statistical Reasoning in the Behavioral Sciences*; *The Male Brain*; *Child Neuropsychology*; *The P Scales*; and on and on. Psychology, I have argued, in its historical roots and in its modern manifestations was and is a cluster of activities with, at most, family resemblance. Sigmund Koch (1917–1996), a doyen of the American profession, observed in 1985: 'After a hundred years of ebullient growth, psychology has achieved a condition at once so fractionated and so ramified as to preclude any two persons agreeing as to its "architecture"'. If, in the mid-twentieth century, the field had American leadership on account of the way quantitative methodology, testing and a public culture of interpreting human problems in psychological terms spread to Europe and the rest of the world, even this did not create unity. U.S. psychology was itself diverse, while elsewhere indigenous traditions re-asserted themselves and new needs shaped new activity.

Nevertheless, there have always been hopes for a unifying scientific theory which, it is said, will enable psychology to emerge as a mature science. Hearnshaw, at the end of a professional life of unusual breadth, published *The Shaping of Modern Psychology* (1987) as a contribution to countering over-specialization and to keeping unity as a goal in view. Since then, there have been advocates of evolutionary neuroscience as the unifying framework. My approach, in contrast, is to understand historically how variety has originated and to think philosophically that variety will continue as long as ways of life vary. Explaining this last point will provide me with a conclusion. First, however, after saying a little more about the growth of the field, I shall focus on three areas of debate: biology and culture, brain and mind and critical thought.

The Second World War repeatedly appears in this story as a watershed. The barbarity of war and of the murderous racial and material aims for which it was fought effected an irreversible moral and political reshaping of the world. Psychology, along with other sciences, came out of the war as part of a worldwide effort to find new ways. The centre of ideological support for the psychological and social sciences was the United States, but as European countries established social democracies and, when countries elsewhere, including Japan, Russia, India, Brazil and China took the road to economic modernization, they also generated conditions promoting psychological activity. Initially, there was optimism about recreating the Enlightenment goal of using reason for human welfare. Later, as gross inequalities and barbarism continued, not least in Europe and in the United States, public disillusion called political support for this into question. By the late twentieth century, though, the institutions of psychology had become deeply entrenched and had their own power and momentum. Moreover, the more the public sphere disappointed, the more, it seemed, people turned to the private world, the world for which psychology had become the natural, everyday language, for what matters in life. Throughout, there was huge commercial interest in recreating ways of life as consumable products, and psychology certainly had its place in this.

The Second World War and the subsequent Cold War familiarized many scientists with the organization of large-scale research, or 'big science'. Psychologists tested all U.S. army officer candidates, for example.

U.S. science policy advisors made it a priority to integrate knowledge and research on human nature and public affairs, and the result was the behavioural sciences. The catalyst was the Ford Foundation, which decided in 1952 to put millions of dollars into academic social science. The Foundation, like the Rockefeller earlier, wanted a co-ordinated strategy not piecemeal, individual research, and it developed the Behavioral Sciences Program as a business plan. The term 'behavior' was taken up because it denoted human activity which could be studied in the manner of the natural sciences and was of immediate social relevance. The intention was to overcome competition between psychology and sociology and create unified science. One concrete result was the establishment in 1952 of the influential Center for Advanced Studies in the Behavioral Sciences at Stanford University in California, an institution which brought together scholars from different disciplines to pursue common themes. Yet, by and large, psychology and sociology remained distinct in the 1950s and 1960s. Later, with the rise to prominence of the neurosciences, behaviour, understood as the expression of nervous and chemical processes serving evolutionary adaptation, was to feature again as a would-be unifying concept.

One consequence of the organization of big science for war and humanity was the rapid expansion of clinical psychology. In the Second World War, large numbers of psychologists gained clinical experience with motivation, stress and injury, and quite a few who had trained with rats or test techniques acquired a taste for work with humans in need of care. The U.S. Veterans Administration, responsible for the wounded, successfully lobbied for federal funds, and this underwrote a large postwar expansion of teaching and employment. An increased number of male as well as female psychologists entered the field, raising its status, and there was a new willingness to accept adults as well as children as clients. The outcome was such a substantial growth of clinical psychology that the APA had to rethink how to accommodate the interests of both academic and applied work. If, earlier, the scientific interest had been unquestionably paramount in professional organization, this was no longer so by 1950.

Psychologists offered knowledge in the form of expertise for the management of mass society – in the classroom, with delinquents, in the army, in industrial enterprises. In the Netherlands, for example, in the 1950s and 1960s the government sponsored psychologists to conduct research on the personality of emigrants, Dutch people who had chosen to make a life elsewhere. Official policy encouraged emigration, and the government therefore sought information to deal with the unwanted 40 per cent return rate within four years. In the course of such work, psychologists achieved relative academic autonomy, created practices independent of medical control and grew as an occupation.

In London, also during the war, the Maudsley Hospital, the leading institution for research in British psychiatry, was set aside for anticipated civilian and military disorders. Here, in 1942, a German immigrant, Hans J. Eysenck (1916–1997), was appointed to head a new sub-department for psychology, which was to become a training ground for clinical work as well as for Eysenck's distinctive use of quantitative tests to classify human personality types and to link them with pathology. Eysenck was a full-time researcher, who never saw patients, consistently interested in using tests and experiments on psychological traits in order to arrive, like Pavlov (an inspirational model), at knowledge of brain processes. His goal was an objective biological science of psychology. His first book (the first of over 80), *Dimensions of Personality* (1947), much praised for establishing quantification in an area notorious for qualitative judgment, developed a two-axis analysis, with a neuroticism dimension and an introvert-extravert dimension. Agreement on personality types proved elusive. Eysenck himself later added a third dimension, while in the 1960s, R. B. Cattell (1905–1998) located sixteen personality factors ('source traits'), fostering a veritable industry of computer-aided tests and correlational analysis. Eysenck's ambition and dynamism also laid the basis for a programme of clinical training and for psychologists to contribute to diagnosis, regulating their own world of testing in the medical context. In the second half of the 1950s, working with the South African psychiatrist Joseph Wolpe (1915–1997), Eysenck introduced behavioural therapy and supplied it with a scientific rationale. This equipped psychologists as therapists, undermining medical monopoly, and it was also part of a self-conscious and aggressive attack on psychodynamic therapies in all their manifestations. Eysenck was a leader of studies which claimed to demonstrate the worthlessness of psychoanalysis. Writing with remarkable self-confidence and speed, he also became the best-known writer of popular psychology in Britain, gaining a large audience for the field.

Eysenck started opposing psychodynamic theories at a time when they had considerable influence within the U.S. psychiatric profession and elsewhere in clinical psychology. Numerous psychologists shared a psychodynamic orientation, while others, particularly those reared, like Eysenck's students, on a diet of testing and statistical analysis, did not. In France, from being the interest of a tiny minority of intellectuals in the 1920s, psychoanalysis had become *the* issue in the organizational struggles of the psychology profession. Psychology was a battlefield between orthodox psychoanalysts, led by Daniel Lagache (1903–1972), and Lacan. Lagache acquired the chair of general psychology at the Sorbonne in 1947 (the first analytically trained psychologist in such a prestigious position), and, in working towards the creation of a national licence for professional psychological work, attempted to balance biological and psychodynamic approaches. It is a history which appears all the more remarkable in the light of the subsequent decline of the status and influence of psychoanalysis in medicine and in psychology. Around 1970, the U.S. psychiatric profession began markedly to turn, or return, to a biological orientation, and even in France Lacan's influence in psychology declined in the following decade.

Testing gave psychologists an occupation which was simultaneously expert and public. It also equipped them with an expertise distinct from the skills needed for experimentation, so much so that Boring noted 'the schism between experimental psychology and mental testing', and then, in 1957, L. J. Cronbach (1916–2001), an educational psychologist interested in measurement, in a presidential address to the APA referred to 'the two disciplines of scientific psychology', general or experimental and differential. As Eysenck's work showed, however, the distinction was not necessarily clear-cut. Moreover, 'general psychology' was itself little more than an umbrella term, including as it did specialities like developmental psychology and physiological psychology. No more than a century earlier was any one version of psychology in a position of undisputed leadership.

Even in the United States, it was only in certain centres where behaviourism appeared the unquestionable framework for scientific psychology, though, admittedly, these centres trained large numbers of students. There was debate about whether research on learning was indeed the route to a unified science or whether it was a label for many different things. Research on memory had long supported other ways of thinking about mind. Thus, about 1970, when it was common to claim that psychology was in the midst of a revolution in which cognitive psychology was supplanting behaviourism, this was a myopic view. Cognitive psychologists investigated perception, problem solving, learning and memory as forms of information processing, and in so doing they referred without embarrassment to internal psychological states. This indeed felt like a revolution to psychologists who had trained to study learning in rats or pigeons and to think of the study of objective variables as intrinsic to science, but the claim that there was a cognitive revolution was an exaggeration. First, non-behaviourist research, like factorial analysis of intelligence, gestalt psychology and Bartlett's work on internal schemata, had always been present: the change overthrew regimes only within particular U.S. institutions. Second, it is a question whether explanatory concepts in fact shifted fundamentally. There was a switch from explanation in terms of behavioural variables to explanation in terms of information-processing variables, but this did not alter the basic assumption that scientific psychology seeks law-like causal, physical explanations. Psychologists who explained human action in other terms did not see a revolution but a change of referents within a natural science world view. Cognitivists, their critics argued, like behaviourists before them, were interested only in understanding being human as individual adaptive reaction to circumstances.

This said, cognitive psychology did become the heartland of academic psychology at the end of the twentieth century. Two reasons were paramount. Psychologists found it necessary to refer to internal psychological states to explain action – the behaviourist attempt to exclude internal variables collapsed. They realized that if they thought of psychological states as *functions* of brain events, even if they then referred to 'internal states', they still referred to natural realities not to illegitimate entities, 'outside nature', like mind or soul. Then there was computing technology: this was something truly revolutionary (if we can apply this misused word to developments lasting decades). Electronic information processing, backed by massive financial and human resources, indeed the commitment of whole national economies, turned cognitive science into an overwhelming influence. The technology created new, powerful methods of analysis; more importantly, it provided an irresistible model of psychological activity – the hardware matches the brain and the software matches the psychological activity.

Interest in cognitive activity began to be apparent in North America in the 1950s, for a number of reasons. It was, for example, one consequence of the use of inferential statistics in the analysis of experimental data. Remarkably, in 1955, more than 80 per cent of experimental articles in the main journals used inferential statistics to justify their conclusions, as this was thought necessary to objectivity. Familiarity with these methods then encouraged psychologists to think analogously about cognitive activity, that is, to talk of mental activity as the analysis of probabilities. It became common to think of people as intuitive statisticians when, for example, they are deciding how to solve a practical problem or remember a face. This pointed towards cognition as a topic and suggested a model of cognition.

A larger influence was renewal of interest in Piaget's work on child development. The imaginative ways in which Piaget conducted his research, using studies of individual children rather than statistics, introduced a breath of fresh air into educational psychology in the 1950s. Piaget attributed development to biologically evolved structures which organize experience and action in a certain sequence as the child grows. Enthusiasm for these ideas brought a structural conception of mind, complete with reference to internal states, into educational psychology, which was a large and influential area of activity. There began to be awareness in the West of the work of Vygotsky, work which similarly referred to stages of development and internal psychological states.

Also during the 1950s, French structuralist theorists questioned, root and branch, the logic and intelligibility of behaviourist and physiological explanation of human language, symbolism and action. This had an influence in the anglophone world later, after psycholinguistics emerged as a field. In North America, one event – Noam Chomsky's review of Skinner's *Verbal Behavior* – became legendary. Chomsky (*b.* 1928), a young professor at MIT, persuaded many people that behaviourist theory could not possibly give an adequate account of language and thus could not pretend to explain human action. Skinner had long recognized language as the test case for his science and, with admirable integrity, developed his work to tackle it. He reportedly never finished reading Chomsky's review, believing Chomsky had completely misunderstood him. Other readers, however, were startled to see a relatively unknown researcher in linguistics demolish a leading psychologist's claims. Psycholinguistics burgeoned as a field, attracting young and talented researchers, and it bypassed issues which had seemed so important to behaviourists. Structuralist thought, drawing ultimately on Saussure, assumed that communication among people is possible, even between people who do not share each other's language, because there are certain formal similarities in all languages, a common structure of language which reflects a common structure of mind. This significantly led psychologists to talk about central, mental conditions.

Chomsky went on to elaborate what he called a 'Cartesian' theory of language, establishing a programme to search out supposed grammatical universals and trace their presence in actual languages. Cartesian linguistics located the base of language in innate mental principles, which Chomsky and his followers thought of, in effect, as the a priori of reason. This strongly stimulated the development of linguistics, though psychologically oriented researchers soon questioned both the logic and the empirical content of the programme, and some even wondered whether Chomsky's early reputation had distorted the development of the field. More empirically based work studied how children do in fact develop language use in different societies. Another reason why Chomsky had an impact was that it was possible to compare what he claimed to be natural grammar with the grammar used in programming.

Mathematicians had stated the logic of the modern computer in the 1930s, and there were also some physical models of calculating machines. The war made it imperative to organize huge systems, like the Manhattan Project and the British air and sea battle against U-boats in the Atlantic. This focused research on encoding, communicating and storing information, along with systems thinking; it also suggested new ways to model human activity. Psychologists were employed to smooth the adaptation of people to machines and also to design machines, especially control systems, which people could actually handle. A mathematician at MIT, Norbert Wiener, suggested unifying principles in a book on *Cybernetics; Or, Control and Communication in the Animal and the Machine* (1948). The research attracted brilliant minds and large funds in both the U.S. and the USSR during the Cold War since weapon delivery systems depended on the efficiency of sophisticated communication and control.

In the late 1930s the English mathematician Alan Turing developed first a theory of computing and then an actual computing machine. Subsequently, he formulated a much-debated criterion in terms of which to judge whether computers 'think': can a person, not in visual contact or otherwise informed, tell, when communicating with a respondent, whether it is a person or a machine? If not, on what grounds can we say a person thinks and the machine does not? In 1943, a paper by W. S. McCulloch (an unorthodox psychiatrist) and W. H. Pitts (a young mathematician), 'A Logical Calculus of the Ideas Immanent in Nervous Activity', noted the similarity between the binary character of logical inference (a logical relation either is or is not) and neural networks in the brain (a neurone fires or it does not). This was part of a move to bring psychological processes into contact with knowledge of brain processes, for which Donald O. Hebb (1904–1985) wrote the most influential statement in *The Organization of Behavior* (1949). The move was hardly possible to carry out at the time, as indeed was recognized at a landmark meeting on the issues, the Hixon Symposium held at Caltech (the California Institute of Technology) in 1948. Through the 1950s, neurophysiology and the study of psychological processes remained largely disconnected from each other. But later researchers were to cite McCulloch and Pitts's paper as having established the arguments which opened the conception of the brain as a computer.

Meanwhile, machines of manageable size, made possible by the substitution of transistors for electronic valves, replaced the large and clumsy computers of the 1940s. Then the microchip began to replace microcircuitry, making possible the mass production and mass marketing of small and cheap computers, the technological basis for a vast transformation of commercial, administrative and private life. This transformation was not just the development of a technology. As psychologists and social scientists began to use information-processing techniques on their data, they began to think about what was going on in the human domain, inside individuals and inside societies, as if this were like, and even as if it were identical with, what goes on inside a processing system. Computing thus became what has been called a 'defining technology', a technology which structures the way people think about the world. It certainly had this effect in psychology: cognitive psychology was the imagination of the computer age applied to knowledge of mind.

New thought in psychological terms about central processes, which in the United States required psychologists self-consciously to reject behaviourist norms, became evident in the late 1950s. Jerome Bruner (*b.* 1915) and George A. Miller (1920–2012), two eminent experimental psychologists at Harvard, established the first cognitive science research centre. Other psychologists were quick to follow once psychologists with a reputation had legitimated reference to central 'plans' in learning and activity. Outside the U.S., in Cambridge in the UK, for example, reference to central psychological processes had never disappeared. By the end of the 1960s, the language of cognition rather than behaviour had become commonplace in many areas of psychology, most conspicuously in fields like perception and memory research.

The shift of attention from behaviour to cognition involved more than a shift of language and a change from the observation of peripheral events to the modelling of central processes. Computing suggested a new way to understand mind's relation to matter. Many people took the obvious step and compared a brain's relation to mind as analogous to a computer's relation to its program. From the beginning, though, there were deep divisions of opinion between scientists who claimed human beings *are* computers (however differently constructed from the machines) and those who drew an analogy. Among the former there was debate about what sort of computer human beings are, while among the latter there was debate about what analogy. There was, of course, nothing new about the claim that man is a machine. What was new was the description of human machinery at the level of *processing* not at the level of mechanical operations.

Creative programming suggested ways in which to model, or perhaps it was even to reproduce, and hence to explain and predict psychological activity. This created the core of cognitive psychology. In the 1960s, the computer scientists Alan Newell and Herbert Simon began to explore how to use computers to solve logical and even scientific research problems, and they hoped that this work in artificial intelligence (AI) would begin to model how humans solve problems. But they did not model the brain. The dominant project in early AI work was the search for a general program for logical problem solving as a model of intelligence. Gradually, however, it became clear that the search for a general program was not plausible as a way to model human activity, and in the 1970s research shifted towards creating different programs for different tasks, encouraging modular theories of mind. The program of 'Deep Blue', the computer which, in 1997 after earlier failures, outplayed Garry Kasparov, the world chess champion, was highly specific, depended on massive computing power and was in no sense a model of the human mind.

If computers intensified the question, 'Is man a machine?', they also posed the question, 'Are computers human?' Some children brought up with computers, for example, wanted to know whether computers could act deliberately and be bad: the children appeared to think that computers pass the Turing test. Philosophers of mind, stimulated by the question why, if computers talk (in principle, if in 2012 poorly in practice) and act, we should not attribute consciousness to them, shifted to thinking in terms of the problem of consciousness rather than the mind–body problem. This became 'the hard problem': understanding how a physical system like the brain could cause conscious 'feels', the qualitative awareness we have of warmth, colour and so on. For the ordinary person, this was the problem of how science could explain the self-evident reality of subjective states and, in grammatical terms, first-person existence. Philosophers divided over whether, and in what way, such problems were conceptual (a matter of how we use language) as opposed to empirical in nature. It became of intense interest to understand mental intentionality, the way the mind, as ordinary language says, acts in response to the *meaning* of information. If 'classical' AI research assumed that thinking involved manipulating formal symbols according to fixed rules, other research assumed that the symbols used in thought denote things in the world and have meaning. Stating the sense in which meaning is part of the world and contributes to psychological life became an outstanding problem.

Increasingly, these issues were tied up with neuroscience. But before turning to neuroscience, I want to return to earlier decades and a debate, with large public participation, which showed that much more was going on under the umbrella of psychology than research on cognition.

## BIOLOGY AND CULTURE

Evolutionary theory argued for continuity between humans and nature. Images of man as monkey pictured this in every Victorian's imagination. But what in fact and in detail was there in common? Belief divided over this from Darwin's age to the present. In recent years, large claims for Darwinian evolutionary theory as the unifying framework of biology and psychology (and even the social sciences) have opposed large and powerful groups, Christian and Islamic alike, rejecting any such approach to the truth.

A reaction against speculative uses of evolutionary ideas to understand past and present human nature set in early in the twentieth century. Separation between the social sciences and biology, leaving psychology awkwardly in the middle, became the orthodox position – of course, with exceptions, not least in the dire instance of racial science – from the inter-war years until the 1970s. It was commonplace to describe the social sciences differing because their subject-matter is 'culture', a word in ordinary language set in opposition to 'nature'. But what is every representation of nature if not the achievement of a historical culture? And what is every aspect of culture if not, in a large sense, part of nature? The conventional division between studies of biology and studies of culture papered over intellectual difficulties which ramified in politics and social policy. Also, the Third Reich brought such horror to the biological language of human differentiation that, for a while, this stifled non-cultural theories of difference. All the same, belief in the hereditary basis of human capacities and the biological basis of human nature remained strong in some professional circles and, it seems likely, in ordinary people's beliefs. Burt and his student Eysenck were notable proponents of the genetic, biological underpinnings of intelligence and personality.

Biological arguments about human nature began to spread again among psychologists and to grip public imagination in the late 1960s. When biology had moved into the laboratory early in the twentieth century, an interest in natural history had persisted, and nobody stopped comparing humans and animals, let alone people and their pets. In the 1940s, a new science of animal behaviour, 'ethology', brought natural history, with its patient ways of observing animals in the field, into contact with university, laboratory-based science. Then, in the 1970s, a number of evolutionary scientists argued for sociobiology, an integration of natural selection theory, ethology and the psychological and social sciences, in order to make the study of people part of biology. Sociobiologists believed that they could achieve unity of knowledge, so conspicuous by its absence in psychology, under the framework of evolutionary theory. It was time, the biologists argued, to reconceive culture as biology. In the 1990s, a second wave of the argument, this time called 'evolutionary psychology', became influential and proved to have substantial popular appeal.

The roots of ethology go back to before 1914. Though laboratory-based research in biology became the norm, individual scientists as well as lovers of natural history continued to seek a less analytic, more direct experience of living nature. The director of the Berlin zoo in the 1920s, Oskar Heinroth (1871–1945), pioneered a critique of a zoo's functions and stressed the differences between wild or natural and captive or artificial animal habits. He wanted to observe what he significantly valued as 'natural' animal behaviour. Heinroth voiced a sensibility distant from the sensibility of animal behaviourists and those who worked on animals in cages and called themselves 'comparative psychologists'. A desire to know 'nature', in humans as in animals, was a powerful theme in an urban and industrial, 'artificial', age. There was a parallel in social psychology when researchers turned to participant observation in 'natural' groups, like work teams, instead of bringing people into the 'artificial' zone of the lab.

Heinroth's colleague, Jakob J. von Uexküll (1864–1944), director of the Hamburg zoo between 1925 and 1944, introduced the concept of the '*Umwelt*', the world to which an animal is bound by its sensory and motor capacities. He thought that the researcher should creatively reconstruct this, the animal's, world. In Utrecht, Buytendijk took up something like a phenomenological approach to the animal mind. His fellow Dutchman Nikolaas Tinbergen (1907–1988) and the Austrian Konrad Lorenz (1903–1989) took such thought further in more rigorous ways. They devised means to observe animal behaviour unaffected by human intervention, refined the concept of instinct and initiated research on inherited behavioural patterns (introducing the notion of imprinting). Tinbergen came to England after a period in a concentration camp and took up a personal chair in zoology at Oxford. There he helped establish a distinct discipline of ethology, a discipline which only subsequently interacted with comparative psychology in the U.S. Lorenz gained a huge audience with his stories of animals, *Er redete mit dem Vieh, den Vögeln und den Fische* (1949; translated as *King Solomon's Ring: New Light on Animal Ways*) and for his studies and beautiful photographs, made over many years, of the greylag goose.

Lorenz's career exposed an ambivalent moral and aesthetic dimension in the critique of what he and many other people felt to be the 'unnatural' character of modern civilization. When he unfavourably compared a loss of instinctual vitality in domestic with wild animals, he also implicitly commented on the human values supposedly lost in modern life. He was repelled by industrial civilization and, perhaps persuaded by rhetoric about the German peoples making a radical return to their roots, he joined the National Socialist or Nazi Party. In 1940, he took an academic appointment in Königsberg and in academic papers linked his biological theories to Nazi ideals, particularly about purifying the *Volk* of degenerate tendencies, though he found little or no support for his work within the Nazi Party. He evidently drew connections between the purity of nature and the purity of human life. In the early 1940s, for a resident of Königsberg, that effectively endorsed Nazi aims in the East.

Much later, in 1963, after time in a Russian camp for prisoners of war, and after years of independent study, Lorenz, in a book on aggression, again commented on human affairs in biological terms. In 1939, he compared the deleterious effects of domesticating animals and people living in large cities. In 1963, he compared aggressive instincts and political activity: 'behavioural science really knows so much about the natural history of aggression that it does become possible to make statements about the causes of much of its malfunctioning in man.' Lorenz's book was joined by a cluster of other studies by authors like Robert Ardrey, Desmond Morris and the opportunely named and more circumspect anthropologists, Robin Fox and Lionel Tiger, which attributed human psychology – aggressiveness, territorial imperatives, emotional expression – to an inherited animal nature. Their work found a receptive public in spite of, or perhaps partly because of, the critical response of social scientists. There appeared to be a deep longing to find a basis for human action beyond politics, a basis in 'real' human nature. The scientists claimed to show through objective observation what was indeed 'natural', and they accused their critics of knee-jerk left-wing politics and the unthinking rejection of biology. The critics accused them in turn of misusing belief in biological determinism to legitimate political inequality and social injustice. There were clear parallels with a debate, going on at the same time, about inheritance and IQ.

Few people were ever so rash as to attribute everything to either nature or nurture. Dispute had rather been about the degree to which it is natural causes rather than social causes which in practice limit human potential and hence the possibilities for change (for example, for reforming criminals). The debate always was political, however much proponents on each side cited empirical evidence, since no one could create knowledge outside a social context. In Britain, for example, there was recurrent conflict over educational policy and resources, with the advancement or retraction of comprehensive education (the same secondary schools for children of all ability in the interest of social equality). No statement about the distribution and causes of intelligence was neutral. It was, however, a paper, 'How Much Can We Boost IQ and Scholastic Achievement?' by Arthur Jensen (1923–2012), in 1969, which brought the question back into the limelight after two decades in which social factors had received sympathetic attention.

In the United States, the question of race and intelligence had again became central to politics in the context of the Civil Rights movement, compensatory educational programmes for black citizens and liberation movements worldwide against political and economic imperialism. Desegregation in education dated from the 1954 Supreme Court decision in *Brown v. Board of Education*, following a hearing for which three psychologists, including the African American Kenneth B. Clark (1914–2005), wrote a paper on the adverse effects of segregation. At the end of the 1960s, the women's movement added a further, highly articulate dimension. Hereditarians, who took race and intelligence to be real biological entities, described objectively by genetic and factorial analysis, portrayed themselves as the scientific vanguard against wishful thinking and policies which could not equalize achievement. It was crucial to judge whether they practised good science.

Jensen argued that 'compensatory education has been tried and apparently it has failed' in the programme for African Americans, called Head Start. He reasserted the central place of general intelligence in human differences, though he did not state differences in racial terms, and, by extension, he argued for the relevance of innate intelligence to social policy. Jensen cited Burt with admiration, and there was a parallel eruption of controversy in Britain. The 1968 Labour government's attempt to advance comprehensive education, which abolished a divide at secondary level between different schools supposedly appropriate to children of different ability, met strong resistance. The temperature of the debate rose after a series of 'Black Papers' argued against the levelling of education. The authors included Burt and Eysenck, and the latter enraged left-wing opinion. But even the controversial Eysenck took second place when a storm broke around the reputation of Burt, who died in 1971 as Britain's most eminent differential psychologist.

Like the pairing of the terms 'nature' and 'nurture', the scientific issue had roots in Galton, in an 1875 comparison of the abilities and character of identical twins. Then and a century later, it appeared that if any evidence would be able to clear up the nature–nurture debate, it would be the evidence of twin studies. Examples of separated identical twins, however, are rare, and researchers had to track them down, compare them with proper controls and study them over an extended period of time. Even when identical twins were separated, they might, for example, have been adopted by similar families, since adopting agencies selected families suitable to adopt children by the same criteria. In spite of these difficulties, however, Burt assembled data on some 30 cases, and his conclusions emphatically supported his view that nature, not nurture, is the key determinant. His results entered the literature and appeared to show scientific psychology producing objective results on a politically sensitive topic.

All this changed. Academics looked into the remarkable precision of Burt's results and into inconsistencies between his papers, and investigative journalism suggested that Burt had not only manipulated data but had invented twins and research assistants. It emerged that Burt's peers well knew that he altered papers in journals he edited to increase the authority of his own point of view. His biographer, Hearnshaw, who in his own career had admired Burt, sadly confirmed the conclusions, though he qualified the stark evidence of fraud with a portrait of Burt's strengths and weaknesses as a psychologist and as a human being. The Council of the British Psychological Society, much divided on the issue, admitted Burt's fraud, though in the eyes of its critics it preferred to denigrate an individual rather than face up to the weaknesses of a discipline which had permitted unacceptable practices. And so the argument went on. Ironically, the dominant view by the 1990s among psychologists was that the cumulative evidence of identical twin studies, based on a long-term U.S. investigation, confirmed a very significant input from heredity into character and ability. Whatever the evidence of genetic endowment, however, numerous biologists and social scientists still pointed out, the endowment always has its actual expression in a social world open to change. According to this viewpoint, it makes no sense and is politically damaging to talk of heredity as if it were autonomous of mode of life.

Independently of developments in psychology, Edward O. Wilson (*b.* 1929) published *Sociobiology: The New Synthesis* (1975), followed by a polemic aimed at the wider public, *On Human Nature* (1978). Wilson was a Harvard zoologist, an authority on the social life of ants, but these books claimed to lay the foundations for a new science, 'the systematic study of the biological basis of all forms of social behavior, in all kinds of organisms, including man', founding ethics, the humanities and the social sciences as well as human biology on a 'truly evolutionary explanation of human behavior'. Wilson detailed the ways in which, he claimed, we can explain, and to a degree predict, human behaviour on the basis of natural selection theory. He picked out four categories of behaviour – aggression, sex, altruism and religion, described them as elemental and attributed them to the inherited survival strategies of a sociable animal. In the light of this, for example, sociobiologists explained the incest taboo as part of a strategy to avoid the deleterious effects of inbreeding, and they explained the practice of women marrying men of greater or at least equal wealth and status as a way to maximize reproductive capacity in hunter-gatherer societies.

To sociobiology's many critics, such writing exhibited disciplinary imperialism and, more profoundly, grossly reduced human existence to one 'real' biological dimension. As Wilson and other sociobiologists wrote for a public audience and signalled the public policy implications of their science, controversy was intense. Feminists were among Wilson's most ardent opponents as the appeal to 'nature' featured so strongly in traditional views of gender divisions. Critics linked sociobiology with the New Right, the backlash against the libertarian politics and lifestyles of the 1960s. For people politically liberal or on the left, sociobiological argument denied the historically and socially constructed character and variability of values and institutions. Conservative writers, for their part, were happy to find support for their belief in the naturalness of fierce individualism, heterosexuality, the family, property, pride in material reward and identification with local community or nation.

Biologists themselves remained divided by Wilson's work and, while sociobiology and biopolitics became established as specialist fields, few scientists were as willing to generalize over such a vast area, and they concentrated instead on detailed studies of animal behaviour, population dynamics and the workings of natural selection. Such research came into focus for psychologists after a long paper by John Tooby and Leda Cosmides, 'The Psychological Foundations of Culture' (1992). This argued for a science of evolutionary psychology, a discipline which would explain human capacities, 'the inherited architecture of the human mind', as the result of natural selection having selected certain types of individuals, because of their gene combinations, during the Pleistocence stage of human evolution. We can learn how the mind works, the authors argued, by understanding the problems of adaptation it was selected to solve in earlier times. A flood of research and, critics thought, fanciful claims in evolutionary psychology followed.

Nowhere was there a closer relationship between animal research and views on human nature than in primatology. Fascination with apes as a mirror to human nature went back beyond the eighteenth century, continued in the Darwin debates – Darwin himself was an attentive visitor to the London zoo – and reappeared in Köhler's experiments with chimpanzees in

Tenerife and in Yerkes's laboratory work, also with chimps. Research and knowledge differed, as ethology and comparative psychology differed, depending on whether it grew in field-based study or laboratory work.

Yerkes's interest, institutionalized in the Yale Laboratories of Primate Biology (opened in Florida in 1930 as the Laboratories of Comparative Psychology), supported intensive studies of chimpanzee learning. The main focus was whether chimps could learn language, since language had standing as *the* capacity unique to humanity. The chimp, Washoe, achieved celebrity status. She was at the centre of a project at the University of Nevada in Reno begun in 1966, and an extended but failed attempt to teach her language rather than the use of signs seemed to put distance between animals and humans. Long before, in 1914–15, the Russian zoologist Nadezhda Ladygina-Kots (1890–1963) had kept a diary of bringing up a male chimp, Ion, and she later compared his progress between eighteen months and four years with the development of her son over the same period. Researchers, however, conspicuously could not agree about results. Subsequently, the publicly acclaimed studies by Jane Goodall (*b.* 1934), who lived for extended periods with chimpanzees in their natural habitat in Gombe in Tanzania, questioned both the intellectual value and ethics of laboratory studies. She was followed into the field by a number of other women, like Dian Fossey (1932–1985) who lived and died (she was, many people believe, murdered) with the Central African mountain gorilla, and they powerfully evoked different values in nature from those which were current in mechanist experimental science. After this work, it became incredible that, in 1935, Yerkes and one of his co-workers had referred to their research on chimp learning as a 'wholly naturalistic study of captive subjects'. The field work, which was a way of life for women who entered with empathy into the world of our nearest animal relatives, created a powerful example of conservationist consciousness, touched perhaps by nostalgia for what people dreamed was humankind's own natural state.

An American feminist historian of science, Donna Haraway, examined the political and cultural dimensions of primate studies, arguing, in *Primate Visions: Gender, Race, and Nature in the World of Modern Science* (1989), that studies mirrored the researchers' own assumptions about human nature. For example, reports about primate sexual and family life included descriptions of male dominance which reflected assumptions about gender identity and roles in contemporary society. Indeed, the word 'gender' came into common currency in the 1960s and 1970s as a way to describe what had conventionally been attributed to differences of sex with a word denoting difference without any assumption about origin. Haraway's book was an extended analysis of how knowledge which appears to come from nature actually comes from social relations mediated through claims about nature. Feminists, all the same, remained divided on the nature–culture issue: there were women who believed that women have a special strength because they have a natural closeness to nature embodied in nurturing qualities. Other women were suspicious of any argument couched in terms of what is natural and instead sought emancipation in a woman's freedom to choose, including the freedom to choose sexual identity.

Arguments for a gendered viewpoint, giving priority to gender in the structure of human experience and affairs, influenced many aspects of the psychological and social sciences. In the mid-1970s, for instance, the philosopher and psychoanalyst Luce Irigaray (*b.* 1932), who was then a colleague of Lacan, asked whether the language in terms of which people speak about femininity presupposes the truth of what they say about it. If so, she argued, it should be possible to reconstruct femininity from a female vantage point in language; this would demystify the feminine and make the feminine, not the masculine, the starting point for speech. Such an argument was intellectually attractive because it suggested ways to reconstruct academic subjects like literary criticism in the light of the new consciousness of gender. The result was to put basic categories for describing human nature, like 'man' and 'woman', into scare quotes: any terms appeared to reflect, as if in distorting mirrors, the language, presumptions and ideology of the speaker. The kind of reflexive play this encouraged was precisely what biologically oriented psychologists, most of all the evolutionary psychologists, intended to replace by the clear truths of natural science.

The issues separating those who emphasized culture rather than biology, or vice versa, were conceptual as well as empirical. This is still so. A large body of opinion has claimed that there is overwhelming evidence to believe in constants in psychological architecture, for example, in colour perception, mechanisms of memory and development in early childhood. These constants, for many psychologists, have become the core subject-matter of the field. Others are not so sure about the evidence. In addition to the empirical questions, however, cultural theorists have argued that there is little gained by reference to psychological constants in the abstract, since a human mind, whatever its architecture, has concrete existence only in the life and language of a particular time and place – in a particular culture. Whatever the biological arguments, understanding the psychology of a person, from this viewpoint, requires knowledge of what a 'person' signifies within a language or symbol system. People exist as living social realities not as abstract biological entities. The social anthropologist Clifford Geertz therefore wrote: 'We are in sum incomplete or unfinished animals who complete or finish ourselves through culture . . . Our ideas, our values, our acts, even our emotions, are, like our nervous system itself, cultural products – products manufactured, indeed, out of tendencies, capacities, and dispositions with which we were born, but manufactured nonetheless.'

## MIND AND BRAIN

By the beginning of the twenty-first century, enthusiasm for an extremely dynamic area of research, the neurosciences, had given these debates a new language and a new focus. The elder George Bush, when U.S. president, declared the 1990s to be 'the decade of the brain'. Claims about what knowledge of the brain would contribute to science, medicine, technology and people's self-understanding were everywhere. This clearly had a profound impact in psychology. A century earlier, psychologists had felt under pressure to say why their field, if it was a natural science, was not simply the branch of biology concerned with the functions of the brain. What was the special subject-matter of psychology? Answers – the conscious mind, behaviour, individual differences of intelligence and personality, child development, cognition, along with the special methods required to study these things – gave scientific psychology autonomy and identity. It was possible on a daily basis to put the question of psychology's relation to physiology, and of mind's relation to brain, to one side. Of course, this generalization must be qualified: the whole point of Pavlov's programme, for instance, was to reveal brain processes. Nevertheless, for much of the century, the question of mind's relation to brain was not a pressing one for scientists, since it appeared that psychology produced more empirical knowledge, and certainly produced more knowledge relevant to application, than research on the brain itself. The brain, as nineteenth-century researchers had found, was formidably difficult to access and formidably complex. By the 1980s, however, the situation had changed, and many scientists were saying that – at last! – progress in brain science would indeed enable psychology to become a true science: neuropsychology. There was large financial support for this belief, sometimes at the expense of other approaches (for example, of observational studies of child development). There was political and public enthusiasm, premised on hopes for medicine but also on a kind of messianic inspiration or faith in knowledge of the brain as the final revelation of truth about the human condition.

The history of this is yet to be written and the outcome of the story is not yet known. I outline the main developments and at the same time suggest there remain tangled questions which empirical advance in neuroscience is not itself going to unravel. Even 'neuroscience' is a misnomer: there are 'neurosciences', in the plural. As the scientist Steven Rose commented: 'we cannot yet bridge the gap between the multiple levels of analysis and explanation, of the different discourses of psychologists and neurophysiologists'.

By the early twentieth century, both psychology and physiology had become large fields, each with specialized sub-disciplines. It was easy for busy researchers to ignore the nebulous philosophical problem of mind's relation to brain. In certain areas, like perception research, psychologists did address the issue; the gestalt psychologists, for example, presupposed isomorphism, parallelism between mental and physical form in the conscious world and in the brain. Philosophers for their part, by and large, considered the mind–brain relation a conceptual rather than empirical matter, a matter of how to use language rather than of facts. Beginning in the late 1970s, however, the situation changed. Many psychologists started to think that knowledge of brain mattered to their work, and an influential number of philosophers of mind proposed, though contentiously, to tackle empirical and philosophical questions in tandem. It became scientific convention to treat mind as a cluster of *brain functions* and thus to seek the causal ground of psychological life in the brain. In polemical rhetoric, science disproved the soul. Francis Crick, earlier a Nobel Prize winner for work on DNA, began a book on the brain with the assertion: '"You," your joys and your sorrows, your memories and your ambitions, your sense of personal identity and free will, are in fact no more than the behavior of a vast assembly of nerve cells and their associated molecules.'

By the beginning of the twenty-first century, brain research had become a huge and prestigious domain. Investment and commitment did not happen just because, as the cliché has it, the brain is the most complex object in the known universe: it is not, since a society, to which many brains contribute, is surely even more complex. Once molecular biology had made possible the understanding of life, and once people had travelled into space, the brain appeared 'the last frontier' of science, the last great challenge to knowledge of humanity's place in the universe. The feeling that this was where the action was going to be attracted scientists into the field. Crick, for example, had moved at the end of the Second World War from physics into molecular biology, and then, after the excitement of discovering the structure of DNA was over, began to search for the material basis of consciousness. There appeared to be a unique depth in research in which the brain attempts to understand itself. (I do not think the brain, as opposed to a person, can do any such thing.) The hope, however, was also for human self-control in an age of potential total self-destruction, hope for knowledge which would enable humanity to transcend the constraint, its own nature, which, many believed, threatened its success. This commonplace was at times a millenarian faith. More mundanely, it legitimated attempts to alter and control the brain, and hence intervene in illness and improve human performance by surgery, drugs, electronic prostheses and genetic manipulation. For advocates of the brain sciences, the new technologies were simply, and brilliantly, the natural development of humanity's long history of extending the sensory and motor reach of the nervous system. They were the next step in evolution. There were also large promises about alleviating mental illnesses (supporting the shift to biological psychiatry) and about securing dignity for, if not halting, suffering and aging.

The modern brain sciences took off in the 1940s under the combined impact of new technologies, new institutional initiatives and war. Before then there was a huge scientific, medical and popular literature on the brain, but, viewed in the light of later developments, it appears speculative and limited. The experimental science of the nervous system, neurophysiology, was highly developed, but the most precise knowledge was about the spinal cord, peripheral nerves and the conduction of the nervous impulse. C. S. Sherrington's *The Integrative Action of the Nervous System* (1906) was the defining text in the English-speaking world, but it said nothing directly about mind. Researchers did not ignore the brain – far from it – but its direct experimental investigation presented outstanding problems. For research on people, with human vivisection unethical, clinical material, so-called natural experiment, was the main source of knowledge.

The situation changed with new technology, especially with electronic recording devices which made it possible to create pictures, representations, of brain activity in living people. First, the thermionic valve led to the oscilloscope (which amplifies an electrical impulse into a visible wave). E. D. Adrian, among others, used this to transform recording techniques in nervous physiology in the 1920s. Then, in 1929, the Jena psychiatrist Hans Berger reported brainwaves, and, after Adrian and Brian Matthews confirmed their existence in 1934, there was a rush to develop knowledge of this expression of brain activity in fully conscious or sleeping subjects. The development of microelectrodes, along with refined surgical procedures for placing them in the brain, made it possible to stimulate and record precisely located neurones. It also allowed for experiments on moving animals and living human subjects. Then, around 1980, the entirely new technology of scanning began to enter the field, with major consequences for psychology. There were to be a number of different kinds of scanning, but they all made it possible to see, with increasing precision, where activity occurs in the brain while a person is carrying out a mental process. In ordinary language, this often, though wrongly, slid into the claim that scanning made it possible to see the mind at work. (What we 'see' is a computer screen with numerical data assembled into a picture which, according to complex transformational rules, records brain events.) The technology shaped neuropsychology. First, the technology worked for humans and did not require experiments involving vivisection with animals which, by the 1980s, had become a public liability for science. Second, it had clear benefits for medical diagnostics. Third, images of brains (or, falsely, minds) at work generated much lay support for neuroscience, for example, via artwork. Fourth, for psychologists it opened a research programme in which it was possible to see progress: the correlation of psychological activity, understood as brain function, with neurone systems.

Technology 'to picture' the mind has been of the greatest importance. It is tempting to think that in brain science, Marshall McLuhan was right: the medium is the message – the technology of representation determines the content of knowledge. This was the case with brainwaves and it was the case again with scanning. History also demonstrates something else: technological complexity leads to a proliferation of specialized sub-disciplines. The result, often enough, is that a scientist's occupational identity comes through training and skill in a technical practice, and this almost inevitably shapes the kind of physical system the brain is thought to be. Yet the path from representation to explanation has proved fraught at best and frustrated at worst. The functional significance of brainwaves remained obscure even after twenty years of enormous investment of time and technological skill in multiple recording techniques. The challenge for scanning research has similarly been to pass from representation to underlying explanation, and this would seem to require agreement about the mind–brain relation. None of this, however, has stopped the representation of brain activity in visual pictures playing a key part in persuading scientists and the public alike that it is indeed possible to understand mental life in physical terms. It has fostered belief in general that mental events *are* physical.

There was also new pharmacological technology. It had long been known that chemical substances, such as curare and strychnine, have specific effects in the nervous system and hence can serve as tools in the analysis of function. Claude Bernard provided models for such research in the mid-nineteenth century, though this was research at the level of the spinal cord. The potential for a chemistry of nervous processes was transformed by extension of research, in the 1930s, on regulation of the heart to demonstrate chemical conduction at synapses, the junctions between neurones. In the early 1950s, the introduction of the drug chloropromazine into huge and hugely overcrowded hospitals for the mentally ill appeared for the first time to suggest a powerful pharmacological answer to insanity. There was also extensive experimentation with LSD, following work by the Swiss researcher Albert Hofmann, published in 1943, and with a host of other preparations. Brain scientists began to take notice in the mid-1960s, after the demonstration of the effects of dopamine on attention and sleep. The more optimistic began to think it possible to locate and determine chemical changes in the brain at the base of processes like memory, sleep and aging. The aspirations of researchers, hopes for therapy, the interests of pharmaceutical companies and popular fascination with altered states of consciousness coincided. A culture familiar with drugs, legal and illegal, developed, bridging scientific and non-scientific communities, sympathetic to the belief that, at base, personality, performance and subjective experience are a matter of chemistry. According to Peter D. Kramer, the author of a bestselling book on the bestselling drug, Prozac: 'Our culture is caught in a frenzy of biological materialism.'

Early financial support came from both the military, especially for but not at all restricted to computing and cybernetics, and medical funders. The Rockefeller Foundation signaled a new interest in the brain and a willingness to commit large-scale resources in the early 1930s. It initially supported work in a piecemeal way, backing, for example, the Kaiser Wilhelm institute for brain research in Berlin headed by Oskar Vogt, where research focused on correlating the micro-anatomy of the visible brain with mental functions. But such work reached a dead end: Vogt sliced Lenin's brain at the urgent behest of the Bolsheviks (the slices are preserved) but even this did not make genius visible. The Rockefeller therefore created a centre which would bring together a range of approaches, medical and experimental, in a unified programme. This centre, opened in 1932, was in Montreal, with the neurosurgeon Wilder Penfield as its director. Penfield carried out famous experiments in which he electrically stimulated areas in the brain in patients under local anaesthetic during exploratory or operative surgery for epilepsy or tumours. Other centres opened, such as John Fulton's at Yale, which standardized the macaque monkey as an experimental subject and extended techniques and units of analysis developed in studies of lower levels to higher levels of the nervous system.

The research of Karl S. Lashley (1890–1958) was an important link between the earlier and later periods and between psychology and neuroscience. He initially worked with Watson and Meyer at the Johns Hopkins University, where he hoped to trace the paths of conditional reflexes in the nervous system. But, turning from this as an unrealizable research project, Lashley challenged the dominant belief in the localization of functions and advanced an alternative theory, the theory of mass action, relating functions such as learning to the whole brain. His empirical evidence came from experiments destroying successively more parts of the rat's brain and then studying what the rat is still able to learn in a maze. As Lashley held a series of chairs in Minnesota, Chicago and Harvard (when he also worked at the Yerkes Laboratories of Primate Biology), he inspired a large number of students with this type of experimental approach to psychological questions. He brought into focus the long-standing divergence of opinion between scientists who wanted to analyse brain function into elementary units and scientists who were committed to a more molar or holistic conception of function.

In 1945 the U.S. government, through the National Institute of Mental Health and the National Institute of Neurological Diseases, committed itself to a concerted effort in the brain sciences, and there were comparable, if smaller, developments elsewhere. Interestingly, this field played a significant role in the gradual integration of scientific institutions into a world community. Brain scientists from the two sides of the Cold War first met each other at a conference in Prague in 1956, when there was informal discussion of Pavlov's legacy. A 1958 colloquium in Moscow, sponsored by UNESCO, brought scientists together on a larger scale and led to the setting up of the International Brain Research Organization (IBRO) in 1960. The work of Luria, in particular, began to be appreciated in the West. The World Federation of Neurology, a forum for physicians, was founded at the same time, in 1957. In the Soviet Union itself, a huge conference on philosophical questions of 'higher nervous activity' and psychology in 1962 was the occasion for a reassessment of Pavlov's legacy, and at this meeting the work of N. A. Bernshtein emerged as the principal alternative way of conceptualizing relations between nervous action and consciousness.

Brain science, unquestionably with U.S. leadership, became one of the fastest growing areas of science during the 1950s, and the pace accelerated in subsequent decades. As noted earlier, however, though experimental physiology of the brain, computing and AI research and experimental psychology all flourished, they did so largely along separate paths. In brain science, for example, there was a specialist field searching for 'the engram', the physical or chemical change imagined to accompany the formation of a memory. It took many years for it to become clear this was not, in general, a profitable research direction, not least because 'memory' may well not be either one thing or any kind of physical storage. Psychological research on memory had meanwhile proceeded on quite different experimental lines. In the field of research on visual perception, by contrast, it did appear possible to use the experimental techniques of physiology to address questions about mental function and even consciousness. Here the early 1960s were a turning point, when it became possible to record nerve impulses from individual cells in the retina and hence to ask questions about the pattern of information processed in the visual cortex of the brain.

The spread of the word 'neuroscience' itself dates from F. O. Schmitt's organization of the Program for Neuroscience at MIT in 1962. Schmitt, a biophysicist, was struck by the rapid progress then being made in a number of areas relevant to the brain but also by the lack of co-ordination and overall strategy. He envisaged neuroscience, in the singular, as the answer, and he brought together an impressive group of researchers from different areas to talk to each other on a regular basis. His programme stimulated communication and promoted an ideal of a unified science. It also had some success: for instance, it brought in the physicist John Hopfield who, in 1982, formulated what became the classic mathematical-computational model of a neural network, a physical system with emergent properties of the kind some people think consciousness might be. (An emergent property is a property, like consciousness, of a complex system, like a brain, which could not have been predicted from the system's elements, the neurones.) The range of activity and the highly specialized technical know-how necessary to be a competent researcher, however, were centrifugal forces, diffusing effort where Schmitt and others hoped to concentrate it. When a sense of shared purpose in brain science and psychology did begin to emerge, there were three reasons. The first was the shift to cognitive approaches in psychology, along with the enormous stimulus of the analogy between computing and cognitive processing. The second was the increased emphasis on evolutionary and genetic studies in psychology, which pointed to an inherited, species-wide material organization, the brain, as the causal foundation of mental life. Lastly, there was new sympathy for empirical approaches to the puzzle of mind's relation to brain among philosophers of mind. A number of psychologists began to feel intense excitement about the field: far from being secondary to other sciences, analysis at the psychological level might turn out to be the key to linking knowledge at the molecular level to knowledge about cultural life. Psychology needed to work 'down' to find connections between psychological and neuronal events, and to work 'up' to find connections between psychological and historical events.

The large body of work in AI and computer modelling of human cognition at first proceeded largely independently of physiological work on the brain. There was 'dry' research on cognition and computers and there was 'wet' research on brain and neurones. The technical competences for each field were very different. The situation began to change in the 1980s, however, when 'dry' cognitive psychologists began to discover 'wet' brain science for themselves. There was much talk about bringing knowledge of brain architecture together with knowledge of cognition to create a science of the functions of actual brains, of natural not artificial intelligence. In the second half of the 1980s, a form of computing known as 'parallel distributed processing', which drives processes at the same time rather than in sequence, caused considerable excitement because it appeared to be much closer to what the brain does. Computers and brains, it was thought, may engage in parallel rather than serial processing, each area of activity linked as part of a connectionist computer or neural net. In the hands of scientists like Gerald Edelman, this turned into a search for the nervous basis of consciousness understood as a brain function carried out by distributed processing. Edelman put forward a notion of consciousness as a continuously changing state of the interaction, or selection, among neurone groups in the brain. Scientists were also excited because this kind of modelling of mind did not require fixed structures.

At the same time as the mind became 'embodied' in neuroscientific research, a number of philosophers of mind turned against the orthodox distinction between conceptual or philosophical and empirical or scientific questions. In 1979, the philosopher John Searle chaired an influential symposium on the relations of neuroscience and philosophy, where a number of scientists *and* philosophers argued that scientific and philosophical questions about mind and brain must be solved together. This promoted a naturalistic theory of knowledge: we can understand what it means to say we know something only in the light of knowledge of brain. There was new enthusiasm for argument about mind just being brain in certain kinds of states; mental activity is simply the functioning of brain. To quote Rose again: 'For many neuroscientists, to ask how the brain works is equivalent to asking how the mind works.' Some of those sympathetic with these arguments thought a revolution in human self-understanding was under way. George Lakoff and Mark Johnson argued: 'What we know about the mind is radically at odds with the major classical philosophical views of what a person is.' Crick trumpeted disproof of the soul. Patricia Smith Churchland and Paul M. Churchland argued that in the future even ordinary people will learn to translate everyday 'folk psychology' into the scientific language of neuronal events. On this view, psychology, as science, will continue as a sub-branch of neuroscience; and we will not talk about emotion but about the amygdala, the mid-brain region where hormones interact with the brain.

Yet a substantial body of philosophical opinion, let alone religious belief, maintained all this was wrong. The Oxford Wittgenstein scholar, P.M.S. Hacker, became something of a spokesman for the argument that there *are* differences between philosophical and scientific questions, and that we do not properly understand mind if we treat it as brain function. Put succinctly, according to this view, using language correctly, we should ascribe mind to *people*, and perhaps elements of mind to animals, not ascribe mind to brains. It also must be said that at the beginning of the new millennium, most researchers, even those sympathetic in principle to the quest, believed no one had any idea how to explain consciousness by the mechanisms of the brain.

Critics have frequently enough attacked brain scientists and physicians for their supposed materialism. Materialism, though it has a long history before the first phrenologists, has not always been prominent in the brain sciences. In Sherrington's generation, and in the generation of scientists and neurologists which succeeded him, which included such people as John Eccles (a Catholic, who worked on the synapse), Penfield, Roger Sperry (who worked on the 'split brain') and the neurologist W. Russell Brain (a Quaker), there was explicit opposition. But the world has changed. Modern 'materialism' originates in the way of life at least as much as in scientific knowledge, as we can see in the word's usage to describe the life of consumption as well as a philosophy of what exists. When people anticipate 'bespoke' pharmaceuticals, drugs designed for individual healthcare and lifestyle, they unite the meanings of the word. Material technology – pharmacology, computing and biological engineering – funded by the interests of capital, is a shaping force in neuropsychology. It all has consequences for what a person is and does.

## THE CRITICAL EDGE

There appears to be a contradiction in psychological society. Psychological thought and practices have flourished in a world of individualistic politics and lifestyle. In ordinary language, the heart of psychological society is the self. Yet psychology has called the very notion of a self into question and instead portrayed the individual as an ephemeral site of material interaction between evolutionary inheritance and circumstance. Moreover, technology is opening the possibility of a person becoming what she or he wishes, or what parents or others in authority have chosen. The computer and the internet will free the mind, surgery and genetic design will free the body and drugs will provide a palette of emotions. The notion of something fixed about being human and belief in a real self has begun to appear antiquated.

If I simplify to make the point, I also exaggerate the role of psychology. Whatever impact it has had in changing notions of the self, this has been as an element in social change – the dissolution of traditional communities, globalization and the flow of capital, chronic insecurity, economic and political migration and the other pressures of the contemporary world. Whatever the reasons, however, psychological society both glorifies and dissolves the self.

Much earlier psychology was concerned with the self, as many practices, like psychotherapy and counselling, still are. This was the guarantee of the humanistic values of the field, the sign that science does not just seek to know but seeks to value the human. For many psychologists, this made psychology intrinsically different from other sciences: whereas science in general sought knowledge of objects, psychology sought knowledge of *subjects*, individual people of value in their own right. When psychology appeared to fail to do this and treated people as if they were objects like anything else, there were critics waiting to pounce, as they did on Skinner's book, *Beyond Freedom and Dignity*. Skinner spoke out loud things which critics thought lurked in all views of human nature grounded on natural science. He promoted his own system of psychology as objective science, central for the human animal's survival strategy. In his terms, survival is a fact not a value; what we *do* is survive. For critics, however, survival is a value – if it were purely a fact, it would not matter whether humans survive or not – and 'mattering', as opposed simply to existing, is the stuff of people's lives. What exists, according to the critics, has meaning as it exists for us, and even when ordinary speech signals lack of meaning (for example, in the phrase 'the meaninglessness of life'), this involves statements with meaning. Behaviourism said nothing about this. If values exist as part of a person's psychological state, and as a matter of fact – it is banal to say it – they do (within the modern frame of consciousness which concerns this argument), then they too are a subject-matter for psychology. The behaviourists, the argument concluded, ignored this and could, therefore, hardly be said to have studied *people*.

Humanistic belief in the essential distinctiveness of the human has had a large place in the field. This was most obviously so in humanistic psychology, which, as an institutionalized movement, was principally the brainchild of Abraham H. Maslow (1908–1970), in the 1950s a professor at Brandeis University in Massachusetts. He established a network of like-minded people and then an association (1963). He and his supporters initially wanted to be a 'third force', an alternative to behaviourism on the one hand and psychoanalysis on the other, which were the two movements then dominant in U.S. psychology. The idea of a new psychology as a 'third force' came originally from G. W. Allport, who had spoken out about the contrast between behaviourist explanations and human aspirations. Humanistic psychology, however, settled down as yet another branch of psychology, with its own section in the APA, and it never achieved the unifying position it had hoped for.

Maslow and his colleagues were committed to a conception of the self, understood as a real entity, as the essential subject-matter of psychology. In this, they followed (even if they were not directly influenced by) the European tradition in philosophical anthropology which sought to state the grounds, in reason and ethics, for the unique worth of being human, for 'the directional and intentional being of man', in the words of the Leiden scholar, C. A. van Peursen. The intellectual problem for psychologists of this persuasion was to formulate a coherent account of the self rather than merely to restate humanistic values, and to come up with concrete proposals for research. Their failure to achieve these ends probably explains the rapid marginalization of their programme. The editorial statement of the first issue of the *Journal of Humanistic Psychology* (founded 1961) well illustrates the problem. The journal was 'interested in those human capacities and potentialities that have no systematic place in either positivistic or behavioristic theory or in classical psychoanalytic theory, e.g. creativity, love, self, objectivity, autonomy, identity, responsibility, psychological health, etc'. This eclectic list consisted of topics marginalized or excluded from psychology rather than topics originating in a coherent theory. Critics found it easy to point both to softness on method and conceptual vagueness. Similarly in Europe, philosophical anthropology came under attack from structuralist thinking in social thought (influenced by Marxism), in linguistics (influencing psychology via Lacan) and in anthropology (with the work of Claude Lévi-Strauss). Then, so-called post-structuralism, led by Michel Foucault, specifically targeted humanism as the evasion of intellectual responsibility: the business of the human sciences should be with the conditions of 'regimes of truth', authoritative ways of discoursing about the relations of knowledge and power, not with claims to the truth of entities like the self.

The historical context of support for humanistic psychology was not natural science but therapy, counselling and existentialist philosophy. It was part of a practical engagement with what people make of life and how they find meaning – preoccupations also targeted by post-structuralists. Precision was not important in the life-world in which humanistic psychology originated, but it became important once the psychology laid claim to be a science. Humanistic psychology was closer to what ordinary people expected of psychology, a practical relationship to life, rather than to what most psychologists thought was required of them as scientists. The popular interest in psychology was an interest in the human condition, coloured by desire for the numinous, as well as in human nature.

Thus humanistic psychology came close to the engagement of psychology with religion. I have touched on the Catholic commitment to a science of the soul, the enthusiasm of Protestant pastors for psychology as the means to speak to the needs of modern congregations and the persistent interest in the psychology of religious experience. But views definitely divided between those who assumed that to explain religious experience psychologically was to 'explain it away' (Freud is the most powerful example) and those for whom psychology was a way to deepen the religious life. Humanists, we might say, wanted the 'deepening', but without the theology.

If early modern *psychologia* was a science of the soul, the dominant psychologies of modern times certainly were not. Somewhere and sometime, if gradually, the language of the soul disappeared and the language of mind replaced it. (I write in English; it is actually not straightforward to state cognates of words like 'soul', 'mind' and 'self' in other languages.) Then, in turn, psychologists replaced reference to 'mind' with the language of 'behaviour' and 'brain function'. All the same, there have been attempts to formulate a modern religious psychology. Some people interpret Jung's psychology in this way. Russia in the post-Soviet period provides an interesting case, since there were attempts to promote an Orthodox psychology, a psychology of the soul based on the Orthodox faith. Given the events in Russia and the demands on people to find new rules by which to live, it is no wonder that some people should have sought to combine ancient and modern, turning to ancient wisdom in the light of modern knowledge. It remained extremely hard, however, to promote the soul as the object of a science in the way that most psychologists understood the notion of a science. It was also difficult to see what concrete research contribution the new religious psychology could make. As the basis for a way of life, however, perhaps it was a different matter.

Humanistic psychology, like humanism generally, took human not divine things to be real, but it shared with religious psychology a sense of the special value, or 'sacredness', of each person. If the language of the soul once denoted this, in English the language of the individual (and 'the rights' of the individual) did so in modern times. Maslow connected the values of U.S. liberal politics, the freedom, dignity and fulfilment of the individual, with the universal, essential needs and goals of the being of individual people. A colleague of Maslow's at Brandeis, the German émigré neurologist Kurt Goldstein (1878–1965), had studied brain damage and disease, during and after the First World War, in terms of the loss of integrity of the whole structure of the brain and whole personality rather than in terms of the loss of specific functions. This holistic orientation, which had intellectual connections with gestalt psychology, gave empirical content to the position of many therapists, European and American. In therapy, it was thought necessary to heal as a response to the potential of a unified self. Many people believed this to be true even for physical medicine. Psychotherapists, for their part, believed that healing required attention to the integrated physical, mental and even spiritual dimensions of the whole person; healing presupposed a capacity for growth and 'self-realization'.

Carl Rogers (1902–1987) gave this kind of practical psychology its most influential form, establishing client-centred therapy, the major idea behind contemporary counselling. His work also illustrates the difficulty of understanding this field as science. Rogers had a religious upbringing among Wisconsin Protestants who stressed the individual soul's or self's relationship with the divine. His beliefs became progressively more liberal when he studied at Union Theological Seminary (associated with Columbia University) in New York, and his interests moved from theology to psychology. This was a not uncommon pattern – Hall, Cattell and Mead were earlier examples – and it explains some of the fervour with which people took up psychology. Rogers moved from the Seminary to Columbia Teachers College and then, in the 1930s, he worked at child guidance, first in New York and then in Rochester in New York State. Caring for children, he entered into extended conflict with the medical profession since he insisted on calling his work 'therapy', although he was not medically qualified. When he moved to the University of Chicago and later to the University of Wisconsin, he applied his approach to mentally ill and distressed adults, and he even took the title of Professor of Psychiatry at Wisconsin, further provoking the physicians. Rogers's opposition to medical monopoly was important because it opened up institutional space in which non-medical therapy subsequently flourished. It gave psychologists access to the huge healthcare market, and this helped medical psychology to expand until it became the largest occupational division within psychology, attracting large numbers of women.

In *Counseling and Psychotherapy* (1942), Rogers described his non-directive stance towards a 'client' – the term he favoured in order to avoid the medical term 'patient' – and the accepting context in which a person can obtain insight into her or his condition. He focused on a person's insight as a cause for change, not, like psychoanalysts, as a route to a hidden past. He presupposed innate goodness or creative power, the self, in every person, though he was not interested in articulating philosophical or religious beliefs. The therapist's or counsellor's role, as Rogers developed it, was to provide acceptance, close to the Christian expression of love, which enables a client to find in herself or himself insight and strength for change. He called this process 'actualization', and, in *On Becoming a Person: A Therapist's View of Psychotherapy* (1961), his ideas reached a large audience. More than most psychologists, Rogers gave therapy or counselling priority, that is, he consciously put theory aside in order to develop an approach through committed practice. While he perhaps did not seek to found a psychological school, his personal qualities as a leader gave him such a position, and he even became a kind of guru for people who wanted from psychology a way of life focused on sensitive and loving openness.

In academic presentations, Rogers referred to 'inner meaning', which he thought pivotal for a person's life, as if it were a psychological variable open to investigation like behavioural variables. He thus rhetorically bridged the academic demand to be objective with the public perception of psychology's task as understanding the private world of meanings. Rogers held out the prospect of an inclusive science of the 'person who creates meaning in life' and thereby has the unity of purpose to live productively. The therapies and practices stemming from or related to his work were legion. They all contributed to a way of life which more and more reconstructed social values as personal feelings and thereby privatized the good and the beautiful. Rogers, and psychotherapy generally, offered hope that this privatized world could still be humane, and they promoted techniques to integrate personal feelings with the feelings of others – an integration significantly called 'interpersonal relations' rather than 'social relations'. Thus political life was carried on in an apolitical way. As for scientific psychology, well, it found here nothing on which to build, though Rogers is credited with persuading psychologists to allow individual descriptions into their work, as opposed to expressing all results as statistical evidence.

One offshoot of Rogers's work was the encounter group movement, a translation of client-centred therapy into a group setting. Encounter groups were intended to create conditions for a person to have insight into his or her relations with others and for this insight to lead to change. The groups were notable for 'the leader's' refusal to lead, forcing participants back on their own creative and, as it could turn out, destructive resources. Rogers was also the major influence on counselling becoming a technique for the management of everyday human problems. During the 1970s and 1980s, counselling entered into all walks of life and, in the activity of churches, merged with pastoral care. All this was central to what I have called psychological society.

Rogers's optimistic, perhaps characteristically American, attitude towards the person was far from the anguished reflections of European existentialists, though both sought to grasp the human situation. European philosophy voiced the horror of mass extermination and the Gulag rather than an optimistic faith in human nature. While the work of philosophers like Husserl and Heidegger was rooted in the technicalities of phenomenology, there were others who found a more accessible language with which to describe the conscious response to a seemingly indifferent world. Some, like Heidegger, eschewed any presuppositions about the base of experience (whether in a god, nature or self) in order to describe the very ground of Being. Others, like the Swiss analyst Ludwig Binswanger (1881–1966), gave their descriptions a psychological character. Responding to the impossibility of faith in God and in moral progress, psychological description turned to anxiety, loss, guilt, meaninglessness and, sometimes, sheer terror. Beginning in the late 1930s, the Paris intellectual Jean-Paul Sartre, who reluctantly accepted the word 'existentialism' to describe his work, combined phenomenological studies of psychological activity like imagination and emotion, as well as working on foundational philosophy – publishing *L'Être et le néant* (1943; *Being and Nothingness*) – and plays and fiction staging harsh human truths.

For Sartre, the core of the human condition is unconditional freedom, the irreducible and primary reality of the actions through which conscious Being becomes what it is. There is, he argued, no self before reflection, and action constitutes a self as a freely chosen existence. It does not on this view make sense to say, for example, that human nature is innately selfish: there is no innate self but, rather, a self as the outcome of free actions, for which, as Sartre uncompromisingly stressed, we are responsible. In this connection, he formulated a much-discussed account of bad faith, a notion which readers easily translated into psychological terms and which, in this form, appeared to have much in common with Freud's notion of unconscious motives. By bad faith, Sartre denoted the myriad ways in which people disguise their choices, hide from themselves and thereby abdicate responsibility for their lives. With a French male example, Sartre described how a woman, seduced by a man, pretends to herself that she is desired for her company and not her physical attractiveness, and she thereby avoids an open choice about a sexual relationship. The woman exhibits bad faith as she does not take responsibility for what she in fact freely chooses. Developed to an extreme, this argument reduced any causal justification for an act, and hence psychology and social science in general insofar as they seek to determine causes, to bad faith. Freedom, it seemed, gives action an ultimate meaninglessness – the state portrayed fictionally by Albert Camus in *L'Étranger* (1942; *The Outsider*). Yet, only by virtue of this freedom, it was thought, is there dignity in being human.

The work of Sartre, Simone de Beauvoir (his lifelong companion) and that of Camus became known in the anglophone world in the late 1940s. In the United States and elsewhere, an artistic avant-garde embraced meaninglessness for its anti-bourgeois message, while others stressed freedom in order to restate each person's dignity and independence. The Protestant theologian Paul Tillich, a socialist refugee from Germany, re-expressed Christian pastoral concern in existentialist language. In his lectures published as *The Courage to Be* (1952), he dealt with the anxiety he found at the centre of experience threatened by nuclear war. He asserted the possibility of an authentic encounter with Being and with a power to overcome anxiety with courage. Tillich was an influential teacher at Union Theological Seminary and later at Harvard, and in these settings psychological and pastoral concern united. Among Tillich's students was Rollo May (1909–1994), a psychoanalyst who published a series of books linking therapy and existentialism. May also co-operated with Maslow, seeking to turn psychology towards humanistic ends, starting from unconditional acceptance of the free person. In Britain, the psychiatrist R. D. Laing (1927–1989) took up existentialist themes and, filtered through his description of schizophrenia as a meaningful response to intolerable family relations, they became a significant element in the anti-psychiatry movement and the 1960s counterculture.

Whatever sympathy academic psychologists had with humane goals, most were sceptical about integrating existentialist or humanistic psychology with science. May, for example, helped create an audience in North America for the Danish Christian philosopher, Søren Kierkegaard, but psychologists found no basis for science in Kierkegaard's account of Abraham's 'fear and trembling' before God. The same could be said about the response to the work of Friedrich Nietzsche. His work dates from the 1870s and 1880s, and his reputation is clearly that of a philosopher. Nevertheless, I want to say something about it. Nietzsche's thought had great influence in the 1960s and 1970s and in certain respects illuminates discontent with psychology as a natural science. This will lead me to conclude by reflecting on the continuing divide between those who think psychology's future lies with it becoming, *tout court*, a natural science (which currently means becoming a neuroscience) and those who think this cannot satisfy human self-knowledge.

Nietzsche made a remarkable prognostication: 'All psychology so far has got stuck in moral prejudices and fears; it has not dared to descend into the depths. [. . . And] the psychologist who thus "makes a sacrifice" [of his own false morality . . .] will at least be entitled to demand in return that psychology shall be recognized again as the queen of the sciences, for whose service and preparation the other sciences exist.' As always with Nietzsche, interpretation is not straightforward. But we can note that 'the queen of the sciences' is a phrase used in medieval times to describe theology. Thus, Nietzsche was saying that future self-knowledge, *if* indeed people 'dare' to seek self-knowledge, lies with overcoming what was previously thought true, or morally right, when we did not understand the sources of thought *in ourselves* (in our 'psychology'). Not reason but 'the will to power', he argued, is the hidden spring of life. This was a philosophical claim, though one easily (if not necessarily correctly) presented in psychological language.

Nietzsche, in his characteristically hyperbolic style, wrote 'that a psychologist without equal speaks from my writings'. This was most obviously a claim about his capacity to go beyond a person's self-description to see hidden motives, to be 'one who has ears behind his ears'. For young people at the beginning of the twentieth century who called themselves 'Nietzscheans', such writing was a liberation from bourgeois superficiality, Christian hypocrisy and academic learning irrelevant to life. There was some sympathy for psychology generally for the same reasons. Nietzsche, in one of his well-known aphorisms, wrote (and Freud, writing in a similar vein, quoted): '"I have done that," says my memory. "I cannot have done that," says my pride, and remains inexorable. Eventually – memory yields.'

Nietzsche used an aphoristic style to compel readers into an attitude of self-questioning and into considering the opposite of what is being said. All this raised large questions for psychology as knowledge. What could be said truthfully about the self if it recreates itself, as in Nietzsche's example, in response to pride and not to memory? What we think we are, it would seem, has come about through self-deception. Is there, then, no real or authentic self but only recreations, using language, of what we persuade ourselves the self is? Scientific psychologists ignored Nietzsche's work, believing their methods gave them the means to be objective – and, within their own terms, this was perhaps so. But Nietzsche's point was that their terms, and even the pursuit of truth and objectivity and moral rightness, are *human* ('psychological') pursuits, the grounds of which we can assert but not show to be objective, true or right.

It was Nietzsche's intent to use language to destroy the clichés of his time – and the capacity of later authors to repeat them. Nevertheless, whatever the use of metaphor, irony and 'the mask' in his writing, he wrote in the late nineteenth century and drew upon biology, the language of 'life', to shape a purifying voice in a world of false religion, politics and psychology of mind. Aspects of this were later to appear highly questionable – his language offered hostages to fortune; what he said about 'woman', for example, now appears consistently appalling. All the same, beginning in the 1960s, readers found in his work a language which located power in every expression and action, not only in the structures of society and the state, and the legacy of such arguments is visible in what observers have said about psychological society. Political power in liberal societies is distributed, diffused and embodied in psychological ways of being and doing. Readers also found a questioning of all fixed foundations for knowledge, and this reading, though hardly touching the natural sciences, did affect the human sciences (psychology included). There was an argument that all possible knowledge of the human, including the humanistic notion of the distinctive value of the human, exists in a historically contingent 'regime of truth', to use Foucault's phrase. In this view, what it is to be a person is not a transcendent truth but a formation of human activity, especially language, which has varied historically and which can change in the future. The argument was a direct attack on the philosophical anthropology of Kant and his successors, including the early Marx, and on the humanistic advocacy of psychological practices. It was also implicitly an attack on the natural science conception of the subject-matter of psychology.

Most scientists, including psychologists, and philosophers too, were suspicious of voices which appeared to question scientific reason and to raise the spectre of the relativity of knowledge. Nietzsche acquired a reputation as the godfather of the late twentieth-century postmodern, fragmented, ironic, eclectic, contradictory and ephemeral culture. Psychologists, by and large, simply ignored this world, except when they thought it attacked science. Intellectuals in the humanities influenced by Nietzsche, Foucault and others lived in one social world, natural scientists in another, and the disciplinary structure of universities made this easy. In these circumstances, however, the human sciences upset the apple cart: they confronted this divorce of convenience with its illegitimacy. Divided within themselves on lines corresponding to the division between the sciences and humanities, both psychology and social science lived with intellectual and practical problems which, though present, other disciplines were not forced to confront.

The rapid expansion of higher education in the 1960s provided institutional support for large communities of both empirically oriented and theoretically oriented academics and students, and for both scientists and radical critics. The pattern was specialization. Yet there were some notable convergences. It was mainstream psychology, not only postmodern theory, which rejected humanistic notions of the self as untenable. Cognitivists and neuropsychologists argued that there is no self, no mind or observer, sitting at the centre of the brain to receive mental images or initiate movements. Marvin Minsky (*b.* 1927), for instance, the leader of AI research at MIT from the late 1960s, criticized ordinary notions of the self as incompatible with science. Neuropharmacology contributed substantially to the postmodern notion of a 'designer self'. Contemporary art frequently took its inspiration from the imagination and technology of the new sciences of the embodied mind. It all pointed to the irony that psychological society accorded the psychological individual unprecedented prominence, yet by the opening of the new millennium, was increasingly explaining the individual as an ephemeral conjunction of material events.

The arrangements suited a lot of people in the sciences and the humanities: they could get on productively with what their discipline had trained them to do, rewarded by their peers. But large questions about what separated the two domains remained; and psychologists had to face up to the way these large questions kept surfacing in the problematic relations between the proponents of psychology as a natural science and their opponents within the field of psychology.

The religious and humanistic response in psychology presupposed a 'sacredness' about the person as an individual (mind-possessing) subject, which contrasts with the character of (physical) objects. This was not necessarily the same as believing in the soul as an entity not explicable by science, and indeed most twentieth-century psychologists were entirely persuaded that natural science forms of explanation are in principle applicable to everything in nature. The point, rather, concerned the fundamental position of values in human life, the necessity for psychology to recognize this, and the absence of values as part of physical nature. Late nineteenth-century German debates provided a reference point, and it became common to distinguish the 'understanding' of a subject with values from the 'causal knowledge' of objects. The difficulty, however, was that Nietzsche and other writers, and then post-structuralist argument, appeared to demolish the intellectual credentials of foundational theories of ethics and of the philosophical anthropology which underpinned humanism. In my view, the natural science advances of psychology, in cognitive science and neuroscience, in the late twentieth century tended in practice towards the same conclusions. Scientific psychologists surely thought of their work as sustaining, even advancing, humane values, but the content of what they claimed about human nature, in principle, rendered values as facts of nature like anything else, with causal origins and a *contingent* nature. The values they propagated in their work derived from the historical culture of which they were part but had no ground *in knowledge*. They dismissed humanistic psychology for its methodological shortcomings, but they achieved their own rigour and objectivity by constructing a form of knowledge which took values as given and no part of their inquiry.

There have been, and are, alternatives. The later philosophy of Wittgenstein, which became well known in the English-speaking world in the 1950s and 1960s, along with analytic philosophy of language, suggested there were reasons to base knowledge of human actions on understanding language rather than causal processes. The impact on psychology was most evident in social psychology, where there was a sense of crisis about lack of direction, since one experimental exemplar, supporting one kind of causal explanation, just seemed to follow another. Rom Harré (*b.* 1927) and P. F. Secord, in *The Explanation of Social Behaviour* (1972), argued that *explanations* of human action require reference to socially situated purposes expressed in the ordinary language of actors themselves. Since people are social beings with language, it is the rules of the social world mediated by language (or other symbol systems) which shape their mental life and what they do: 'The idea of men as conscious social actors, capable of controlling their performances and commenting intelligibly upon them, is more scientific than the traditional conception of the human "automaton".' Harré went on to develop this into discursive psychology, the analysis of people's statements about life and actions. To ask what someone feels or why she does something is in ordinary speech to ask about the *meaning* of mental events and action in the course of life. Causal knowledge (such as about neurones), even were it to exist in detail, would not answer these questions. While natural scientists think it in principle possible to explain everything, including the human world, in material causal terms, when we ask for an explanation of human action we normally ask for something different. We seek to know about the intentions and reasons for the actions, which exist as part of collective life and broadly follow social rules or customs. It would be odd to say that you read this book because your neurones fire in one way or another (though they do), and it would be simply wrong to say that neurones, rather than you, do the reading. You read because you have purposes relating to training, occupation, beliefs and preferences. As an individual, you have a plan to carry out, and the plan, however individually coloured, conforms to complex social conventions. (A large debate followed about whether the reasons people have should be construed as causes: there were theorists who thought social psychology should become a causal but social science, and those who maintained that psychological understanding is not causal knowledge at all. There were many positions.)

A number of possibilities opened up. Particularly interesting for a historian was a new evaluation of history's relation to psychological knowledge: if reasons for actions lie in (linguistically mediated) ways of life, then the history of ways of life becomes fundamental to psychology. Kenneth J. Gergen (*b.* 1935) therefore reconceived social-psychological knowledge as historical knowledge and, in his paper on 'Social Psychology as History' (1973), argued that social psychologists study a historically specific subject-matter and not general actions, not 'nature': 'social psychological research is primarily the systematic study of contemporary history.' Gergen and others, drawing in the 1970s on the field of sociology of knowledge, elaborated social constructivism, arguing that we must understand psychological events as social action rather than as processes based on 'natural' realities pre-existing society. The German-born Canadian social psychologist and historian of psychology, Kurt Danziger (*b.* 1926), in *Constructing the Subject* (1990), described in detail how experimental social psychologists constructed their subject-matter, such as personality, in the social relations which they set up in their experiments. This history, he implied, *is* social psychology. He subsequently expanded his work to suggest historical roots for the construction of the very categories which psychologists take for granted – treat as 'natural kinds' not 'human kinds', given by nature and not formed by human intervention – like memory, intelligence and emotion.

Some psychologists and historians of psychology argued that the *history* of subjectivity and of psychological states, like emotions, is of central importance to psychology, and they looked to disciplines in the humanities, literary studies and art history, for knowledge about people's historical and psychological nature. This took up the tradition of historical psychology mentioned in connection with social psychology. At the back of such work was a double question: have people since ancient times had psychological categories, implicitly if not explicitly, and have these categories denoted the same subject-matter? That is, have people always made use of words like 'memory', 'emotion', 'intelligence', 'behaviour' and so on, and have these words had a constant denotation? Of course, yes, more or less, I imagine most Westerners would now say: the ancient Greeks had emotions and they had a language for the emotions. But so unquestioning a stance takes for granted what actually requires to be shown. It is highly complicated. First, and obviously, we must judge the translatability of words and concepts, such as *psuchē*, *anima* and *Geist*. Then, we need to decide if it is meaningful to say the ancients had psychological states, though they had no word for psychology and no conception of such a field. More specifically, is it historically correct to equate passion and emotion, reason and intelligence, character and personality and melancholia and depression? Though modern usage may treat a passion as merely a strong emotion, it was once the condition of a soul as suffering object (as, indeed, in Christ's Passion); memory in Platonic thought was a state of the rational soul which it was the true end of life to recover by virtuous living, not a computer store; and talk about 'conduct' deployed a moral category, while talk about 'behaviour' referred to a biological one.

There has been a critical, political edge to these questions. Learning about the historical formation of psychology's subject-matter is a way to understand the historical formation of the occupations of psychologist. History therefore asks psychologists to act reflexively, to include on their agenda the social analysis of the context and historical roots of their own individual and collective work. Obviously, it is not only history which asks this; ethical issues (raised, for example, by Milgram's experiments on obedience) and black and feminist consciousness have done the same.

In the 1970s, small but intense groups of psychologists, led by Klaus Holzkamp (1927–1995) in Berlin and K. F. Riegel in North America, attempted to give psychology a Marxian framework. Even when Marxian terms went into decline, observers still discussed how free market ideology promoted the naturalness of thinking about society as a collection of psychological individuals rather than seeing the notion of *Homo psychologicus* as itself the historical creation of a way of life. The last two decades of the twentieth century saw a marked shift in moral and political commitment, especially in the United States but followed elsewhere, stressing the individual and innate determinants of economic and social achievement (or failure). Accounts of the evolutionary process and of the political process became remarkably similar: writers pictured events in both domains as the competition between biologically defined individuals, and in more extreme accounts it was not even individuals who competed but genes. Thus, the new millennium began with widespread openness to belief that the foundation of human self-understanding lies in the biology of inheritance. To those sympathetic to this position, brain research appeared the avenue by which the general belief would become detailed knowledge. For critics, neither brain research nor evolutionary theory, however exciting as causal knowledge, was a source of understanding of human mental, linguistic and social forms of life – including the forms of life of psychologists themselves *as historical actors*. In this context, psychologists refer to critical psychology, using the word 'critical' to signal a position 'outside' psychology which creates perspective on what psychology is and does. My history, I hope, is a contribution.

The domain of psychology was never an empty land waiting for psychologists to march in to claim their kingdom. Psychologists and ordinary people, who are psychologists too, created this land: it was a historical achievement. As I have told the story, the story is modern, indeed central to what defines modern life as 'modern'. As Nietzsche anticipated, though the results are hardly what he would have wished, the turn to psychology is a turn to a modern form of truth. Beginning in the Enlightenment, running through the high tide of nineteenth-century belief in progress and continuing into the present, the quest has been to learn what it is natural for people to do, to study their 'psychology', in order to live well. Yet as Tolstoy argued, echoed famously by Weber, science could not provide answers about how to live. To 'the only important question: "What should we do? How should we live?"', Weber said, 'the fact that science does not give an answer is completely undeniable.' Countering this, it is true, there are evolutionary theorists who say that human values, like human capacities, originate in nature, and that certain values are therefore natural and thereby legitimate (though, as Diderot long ago quipped, this would seem to make being nasty as natural as being nice). Others turn to divine revelation. Yet other people hold that values are an expression of language and exist only where history makes them possible. Many scientists themselves tacitly suppose this when they hold up the *way of life* of science, with its ideal openness to objectivity, as a model for society generally.

Psychologists have taken, and continue to take, many positions, and the field is a divided one. This, I would think, is a cause for celebration, not distress: it is bureaucracy not truth which dictates uniformity in forms of life.

The enormous endeavour in different areas of psychology did not produce a synthesis with which to start the twenty-first century. Some psychologists hoped study of the brain would create unity, as earlier psychologists had hoped the study of behaviour and others the study of cognition would do. Though the goal of rational knowledge required psychology to become a science in some sense, the record is of people thinking about science in different ways. The split remains between those convinced psychology is a natural science and those who seek other knowledge. There are large numbers of people who do not much worry about science but hope psychology will provide answers to how to live, and there are many others who think the whole direction of contemporary scientific psychology misguided. It is easy to be bewildered. Thus, history makes a contribution. It makes clear the stories to hand for understanding a place in the world.

## READING

I take information about the scale of psychological activity from such sources as E. R. Hilgard, *Psychology in America: An Historical Survey* (San Diego, 1987), and A. R. Gilgen and C. K. Gilgen, *International Handbook of Psychology* (New York, 1987). More recent work, sensitive to the world dimension, includes A. C. Brock, ed., *Internationalizing the History of Psychology* (New York, 2006), and D. B. Baker, ed., *The Oxford Handbook of the History of Psychology* (New York, 2012). For the political culture of psychology in the U.S., E. Herman, *The Romance of American Psychology: Political Culture in the Age of Experts* (Berkeley, CA, 1995). There is a fine biography of Eysenck, full of information about postwar British psychology: R. D. Buchanan, *Playing with Fire: The Controversial Career of Hans J. Eysenck* (Oxford, 2010). For cognitive psychology, H. Gardner, *The Mind's New Science: A History of the Cognitive Revolution* (New York, 1985), and for the enthusiastic, M. A. Boden, *Mind as Machine: A History of Cognitive Science*, 2 vols (Oxford, 2006).

Thought about biology and society is surveyed in C. N. Degler, *In Search of Human Nature: The Decline and Revival of Darwinism in American Social Thought* (New York, 1991). For Lorenz: T. J. Kalikow, 'Konrad Lorenz's Ethological Theory: Explanation and Ideology, 1938–1943', *Journal of the History of Biology*, XVI (1983), pp. 39–73, and 'Konrad Lorenz's "Brown Past": A Reply to Alec Nisbett', *Journal of the History of the Behavioral*

*Sciences*, XIV (1978), pp. 173–80. There is a critique of biologized views of society in S. Rose, L. J. Kamin and R. C. Lewontin, *Not in Our Genes: Biology, Ideology, and Human Nature* (New York, 1984). For balanced discussions: A. Kuper, *The Chosen Primate: Human Nature and Cultural Diversity* (Cambridge, MA, 1994); K. Malik, *Man, Beast and Zombie: What Science Can and Cannot Tell Us about Human Nature* (London, 2000); and M. Teich, R. Porter and B. Gustafsson, eds, *Nature and Society in Historical Context* (Cambridge, 1997). For an anthropological perspective on psychology as a social project rather than as part of a natural order of knowledge, M. Sahlins, *The Western Illusion of Human Nature: With Reflections on the Long History of Hierarchy, Equality, and the Sublimation of Anarchy in the West, and Comparative Notes on Other Conceptions of the Human Condition* (Chicago, 2008).

Modern neuroscience largely awaits its historians. But there are some detailed studies, including S. J. Heims, *The Cybernetics Group* (Cambridge, MA, 1991), and N. M. Weidman, *Constructing Scientific Psychology: Karl Lashley's Mind–Brain Debates* (Cambridge, 1999). On mind–body theories from the 1950s to the 1980s, S. Moravia, *The Enigma of the Mind: The Mind–Body Problem in Contemporary Thought*, trans. S. Staton (Cambridge, 1995). For recent activity, there are accessible guides, like S. Rose, *The 21st-Century Brain: Explaining, Mending and Manipulating the Mind* (London, 2006). The philosophical arguments critical of understanding mind as a function of brain are laid out in M. R. Bennett and P.M.S. Hacker, *Philosophical Foundations of Neuroscience* (Malden, MA, and Oxford, 2003). A precise statement of the sociological understanding of mental life and human action is J. Coulter, *Mind in Action* (Atlantic Highlands, NJ, 1989); also M. Kusch, *Psychological Knowledge: A Social History and Philosophy* (London, 1999).

Psychologists have on a number of occasions set up dialogue between natural science and phenomenology, as in T. W. Wann, ed., *Behaviorism and Phenomenology: Contrasting Roots for Modern Psychology* (Chicago, 1964), and S. B. Messer, L. A. Sass and R. L. Woolfolk, eds, *Hermeneutics and Psychological Theory: Interpretive Perspectives on Personality, Psychotherapy, and Psychopathology* (New Brunswick, NJ, 1990). An interesting model for combining science and the humanities is J. Starobinski, *Action and Reaction: The Life and Adventures of a Couple*, trans. S. Hawkes (New York, 2003). On the humanists: R. J. DeCarvalho, *The Founders of Humanistic Psychology* (New York, 1991); and for the 'critical psychologists', T. Teo, *The Critique of Psychology: From Kant to Postcolonial Theory* (New York, 2005). K. Danziger (in addition to the work cited in reading for chapter Four) elaborated on the historical origin of psychological categories in *Marking the Mind: A History of Memory* (Cambridge, 2008); see also D. Draaisma, *Metaphors of Memory: A History of Ideas about the Mind*, trans. P. Vincent (Cambridge, 2000). On the category of emotion, D. M. Gross, *The Secret History of Emotion: From Aristotle's 'Rhetoric' to Modern Brain Science* (Chicago, 2006); T. Dixon, *From Passions to Emotions: The Creation of a Secular Psychological Category* (Cambridge, 2003). Ian Hacking's work, especially *Rewriting the Soul: Multiple Personality and the Sciences of Memory* (Princeton, NJ, 1995), spread the notion of a 'historical ontology' of psychological categories. For a concise study relating cross-cultural psychology to the anthropological and historical evidence, G.E.R. Lloyd, *Cognitive Variations: Reflections on the Unity and Diversity of the Human Mind* (Oxford, 2007).

# REFERENCES FOR QUOTATIONS

## ONE: EARLY STRANDS OF MIND


p. 9 M. Merleau-Ponty, *Phenomenology of Perception*, trans. C. Smith (London, 2002), p. xxii.

p. 12 F. Nietzsche, *Beyond Good and Evil: Prelude to a Philosophy of the Future*, trans. W. Kaufmann (New York, 1966), section 23.

p.13 G. Ryle, *The Concept of Mind* (Harmondsworth, 1963), p. 305.

p.13 M. de Certeau, *The Practice of Everyday Life*, trans. T. Conley (Berkeley, CA, 1984), p. 50.

p. 15 Barry, quoted in G. Richards, 'Psychology and the Churches in Britain 1919–39: Symptoms of Conversion', *History of the Human Sciences*, XIII/2 (2000), p. 64.

p. 16 I. Hacking, 'The Looping Effects of Human Kinds', in *Causal Cognition: A Multidisciplinary Debate*, ed. D. Sperber, D. Premack and A. J. Premack (Oxford, 1995), pp. 351–94.

p. 16 M. Foucault, *The Order of Things: An Archaeology of the Human Sciences*, trans. A. Sheridan (London, 1970), p. 327.

p. 26 de Mably, quoted in L. G. Crocker, *Nature and Culture: Ethical Thought in the French Enlightenment* (Baltimore, ML, 1963), p. 480.

p. 27 Locke, quoted in C. Fox, *Locke and the Scriblerians: Identity and Consciousness in Early Eighteenth-Century Britain* (Berkeley, CA, 1988), p. 1.

p. 28 Bevington, quoted in R. Lekachman, *A History of Economic Ideas* (New York, 1964), p. 104.

p. 32 A. Smith, *The Theory of Moral Sentiments*, ed. D. D. Raphael and A. L. Macfie (Indianapolis, 1984), p. 110.


## TWO: THE MIND'S PLACE IN NATURE


p. 39 I. Turgenev, *Fathers and Sons*, trans. R. Edmonds (London, 1975), p. 90.

p. 41 Gall, quoted in R. M. Young, *Mind, Brain, and Adaptation in the Nineteenth Century: Cerebral Localization and Its Biological Context from Gall to Ferrier* (Oxford, 1970), p. 12.

p. 42 L. von Ranke, literary remains, in F. Stern, *The Varieties of History: From Voltaire to the Present* (New York, 1973), p. 61.

p. 43 T. Laycock, *Mind and Brain: Or the Correlations of Consciousness and Organization* (Edinburgh, 1860), vol. I, p. 1.

p. 43 A. Bain, *The Senses and the Intellect* (London, 1855), p. v.

p. 47 Moleschott, quoted in F. Gregory, *Scientific Materialism in Nineteenth Century Germany* (Dordrecht, 1977), pp. 89–90.

p. 49 A. Bain, 'The Respective Spheres and Mutual Helps of Introspection and Psycho-Physical Experiment in Psychology', *Mind*, new series 2 (1893), p. 42.

p. 50 D. Ferrier, *The Functions of the Brain* (London, 1876), p. 275.

p. 51 T. H. Huxley, *Man's Place in Nature, Collected Essays*, vol. VII (London, 1894).

p. 53 Hall, quoted in S. J. Gould, *The Mismeasure of Man* (Harmondsworth, 1984), pp. 116–17.

p. 54 C. Darwin, letter to Henry Fawcett, 1861, in *The Correspondence of Charles Darwin, Volume 9: 1861* (Cambridge, 1994), p. 269.

p. 55 C. Darwin, *The Descent of Man, and Selection in Relation to Sex* (Princeton, NJ, 1981), vol. II, p. 404.

p. 55 C. Darwin, *On the Origin of Species by Means of Natural Selection*, ed. J. W. Burrow (Harmondsworth, 1968), p. 458.

p. 56 Darwin, *Descent*, vol. I, p. 71.

p. 57 A. R. Wallace, 'The Origin of Human Races and the Antiquity of Man Deduced from the Theory of "Natural Selection"', *Journal of the Anthropological Society of London*, II (1864), p. clxviii.

p. 58 C. Darwin, *The Expression of the Emotions in Man and Animals* (Chicago, 1965), p. 19.

p. 60 Reported by Spencer, quoted in Young, *Mind, Brain, and Adaptation*, p. 151.

p. 60 H. Spencer, *First Principles* (London, 1915), p. 321.

p. 60 W. James, 'Herbert Spencer's Autobiography', in *Memories and Studies* (London, 1911), p. 124.

p. 61 H. Spencer, *The Principles of Psychology* (London, 1855), p. 606.

p. 63 T. Ribot, *German Psychology of To-Day: The Empirical School*, trans. J. M. Baldwin (New York, 1886), p. 5.

p. 64 Morgan, quoted in A. Costall, 'How Lloyd Morgan's Canon Backfired', *Journal of the History of the Behavioral Sciences*, XXIX (1993), p. 116.

p. 66 T. H. Huxley, 'On the Hypothesis that Animals Are Automata and Its History', in *Method and Results: Essays* (London, 1894), p. 242.

p. 66 W. James, 'Remarks on Spencer's Definition of Mind as Correspondence', in *Collected Essays and Reviews* (London, 1920), p. 67.

### THREE: SHAPING PSYCHOLOGY

p. 70 W. James, *The Principles of Psychology* (New York, 1950), vol. I, p. 402.

p. 73 T. Ribot, *English Psychology*, trans. J. Fitzgerald (London, 1873), p. 25.

p. 74 T. Ribot, *The Diseases of Personality*, trans. anon. (Chicago, 1906), pp. 1–2.

p. 76 Mercier, quoted in H. Misiak and V. M. Staudt, *Catholics in Psychology: A Historical Survey* (New York, 1954), pp. 44 and 47.

p. 86 Ebbinghaus, quoted in E. G. Boring, *A History of Experimental Psychology* (New York, 1950), p. 392.

p. 87 E. Mach, *The Analysis of Sensations and the Relation of the Physical to the Psychical*, trans. C. M. Williams, revd S. Waterlow (New York, 1959), p. 310.

p. 89 Stumpf, quoted in M. G. Ash, 'Academic Politics in the History of Science: Experimental Psychology in Germany, 1879–1941', *Central European History*, XIII (1980), p. 269.

p. 91 W. Dilthey, *Descriptive Psychology and Historical Understanding*, trans. R. M. Zaner and K. L. Heiges (The Hague, 1977), p. 27.

p. 95 Hall, quoted in J. M. O'Donnell, *The Origins of Behaviorism: American Psychology, 1870–1920* (New York, 1985), pp. 119 and 128.

p. 96 Scripture, quoted in O'Donnell, *Origins of Behaviorism*, p. 38.

p. 97 W. James, 'A Plea for Psychology as a "Natural Science"', *Philosophical Review*, I (1892), p. 146.

p. 100 Pillsbury, quoted in O'Donnell, *Origins of Behaviorism*, p. 131.

## FOUR: PSYCHOLOGICAL SOCIETY

p. 102 C. K. Ogden, *The ABC of Psychology* (London, 1929), p. 1.

p. 103 Quoted from the founding meeting of the BPS in S. [A. D.] Lovie, 'Three Steps to Heaven: How the British Psychological Society Attained Its Place in the Sun', in *Psychology in Britain: Historical Essays and Personal Reflections*, ed. G. C. Bunn, A. D. Lovie and G. D. Richards (Leicester, 2001), p. 97.

p. 108 L.-A. Bertillon, quoted in J. Cole, 'The Chaos of Particular Facts: Statistics, Medicine and the Social Body in Early 19th-Century France', *History of the Human Sciences*, VII/3 (1994), p. 20.

p. 110 The Spens Report, quoted in B. Simon, *The Politics of Educational Reform, 1920–1940* (London, 1974), pp. 249–50.

p. 112 A. Binet and V. Henri, 'La psychologie individuelle', *L'Année psychologique*, II (1895), p. 411.

p. 113 Stern, quoted in R. E. Fancher, *The Intelligence Men: Makers of the IQ Controversy* (New York, 1985), p. 101.

p. 114 Terman, quoted in H. L. Minton, 'Lewis M. Terman and Mental Testing: In Search of the Democratic Ideal', in *Psychological Testing and American Society*, ed. M. M. Sokal (New Brunswick, NJ, 1987), p. 100.

p. 116 E. G. Boring, 'Intelligence as the Tests Test It', in *History, Psychology, and Science: Selected Papers*, ed. R. I. Watson and D. T. Campbell (New York, 1963), p. 18.

p. 117 Thorndike, quoted in K. Danziger, *Constructing the Subject: Historical Origins of Psychological Research* (Cambridge, 1990), p. 235.

p. 118 Maine de Biran, *De l'aperception immédiate: Mémoire de Berlin 1807*, ed. A. Devarieux (Paris, 2005), p. 107.

p. 122 M. Mead, quoted in C. N. Degler, *In Search of Human Nature: The Decline and Revival of Darwinism in American Social Thought* (New York, 1991), p. 134.

p. 122 M. Mead, quoted in Degler, *In Search of Human Nature*, p. 135.

p. 122 A. T. Poffenberger, quoted in D. S. Napoli, *Architects of Adjustment: The History of the Psychological Profession in the United States* (Port Washington, NY, 1981), p. 30.

p. 125 Prospectus of Tavistock Clinic, quoted in L. S. Hearnshaw, *A Short History of British Psychology, 1840–1940* (London, 1964), p. 284.

FIVE: VARIETIES OF SCIENCE

p. 137 J. B. Watson, *Psychology from the Standpoint of a Behaviorist* (London, 1983), p. 3.

p. 137 J. S. Mill, 'Bain's Psychology', *Edinburgh Review*, CX (1859), p. 287.

p. 137 James, *Principles of Psychology*, vol. I, p. 1.

p. 137 J. Ward, *Psychological Principles* (Cambridge, 1918), p. 104.

p. 137 J. B. Watson, 'Psychology as the Behaviorist Views It', *Psychological Review*, XX (1913), p. 158.

p. 138 S. Koch, 'The Nature and Limits of Psychological Knowledge: Lessons of a Century qua "Science"', in *A Century of Psychology as Science*, ed. S. Koch and D. E. Leary (New York, 1985), pp. 92–3.

p. 139 F. Bartlett, *Thinking: An Experimental and Social Study* (London, 1958), pp. 132–3.

p. 141 Watson, 'Psychology as the Behaviorist Views It', p. 163.

p. 142 Watson, *Psychology from the Standpoint of a Behaviorist*, pp. 364–5.

p. 143 Thorndike, quoted in G. W. Allport, 'The Historical Background of Modern Social Psychology', in *The Handbook of Social Psychology*, ed. G. Lindzey and E. Aronson (Reading, MA, 1968), vol. I, p. 15.

p. 144 Quoted in D. Bakan, 'Behaviorism and American Urbanization', *Journal of the History of the Behavioral Sciences*, II (1966), p. 6.

p. 148 B. F. Skinner, 'Behaviorism at Fifty', in *Behaviorism and Phenomenology: Contrasting Bases for Modern Psychology*, ed. T. W. Wann (Chicago, 1964), p. 88.

p. 150 I. P. Pavlov, *Lectures on Conditioned Reflexes: Twenty-Five Years of Objective Study of the Higher Nervous Activity (Behaviour) of Animals*, trans. and ed. W. H. Gantt (London, 1963), vol. I, pp. 38–9.

p. 152 Pavlov, quoted in D. Joravsky, *Russian Psychology: A Critical History* (Oxford, 1989), p. 134.

p. 153 Bukharin, quoted in Joravsky, *Russian Psychology*, p. 253.

p. 156 Husserl, quoted in M. G. Ash, 'Gestalt Psychology: Origins in Germany and Reception in the United States', in *Points of View in the Modern History of Psychology*, ed. C. E. Buxton (Orlando, FL, 1985), p. 301.

p. 157 Katz, quoted in H. Spiegelberg, *Phenomenology in Psychology and Psychiatry: A Historical Introduction* (Evanston, IL, 1972), p. 47.

p. 158 Buytendijk, quoted in T. Dehue, *Changing the Rules: Psychology in The*

*Netherlands, 1900–1985* (Cambridge, 1995), p. 69.
p. 158 F.J.J. Buytendijk, 'Husserl's Phenomenology and Its Significance for Contemporary Psychology', trans. D. O'Connor, in *Readings in Existential Phenomenology*, ed. N. Lawrence and D. O'Connor (Englewood Cliffs, NJ, 1967), p. 353.
p. 159 Merleau-Ponty, *Phenomenology of Perception*, p. 9.
p. 161 Koffka, quoted in Ash, 'Gestalt Psychology', p. 310.

## SIX: UNCONSCIOUS MIND

p. 171 R. Barthes, *A Lover's Discourse: Fragments*, trans. R. Howard (London, 1990), pp. 229–30.
p. 172 S. Freud, *The Interpretation of Dreams*, in *The Standard Edition of the Complete Psychological Works of Sigmund Freud*, ed. J. Strachey (London, 1953–74), vol. V, p. 613.
p. 173 S. Freud, letter to W. Fliess, 1900, in *The Complete Letters of Sigmund Freud to Wilhelm Fliess, 1887–1904*, ed. J. M. Masson (Cambridge, MA, 1985), p. 398.
p. 178 J. Breuer and S. Freud, 'Preliminary Communication', in *Studies on Hysteria, Standard Edition*, vol. II, p. 7.
p. 178 S. Freud, *The Interpretation of Dreams, Standard Edition*, vol. IV, p. 101.
p. 180 J. Breuer, in *Studies on Hysteria, Standard Edition*, vol. II, p. 246.
p. 183 S. Freud, *Group Psychology and the Analysis of the Ego, Standard Edition*, vol. XVIII, p. 92.
p. 183 S. Freud, *Beyond the Pleasure Principle, Standard Edition*, vol. XVIII, p. 38.
p. 184 S. Freud, in *Studies on Hysteria, Standard Edition*, vol. II, p. 305.
p. 187 J. Lacan, *Écrits: The First Complete Edition in English*, trans. B. Fink, with H. Fink and R. Grigg (New York, 2006), p. 6.
p. 188 P. Ricoeur, *Freud and Philosophy: An Essay on Interpretation*, trans. D. Savage (New Haven, CT, 1970), pp. 32–6 and 529.
p. 189 S. Freud, *New Introductory Lectures on Psycho-Analysis, Standard Edition*, vol. XXII, p. 126.
p. 189 S. de Beauvoir, *Nature of the Second Sex*, trans. H. M. Parshley (London, 1963), p. 8.
p. 192 Quoted in J. C. Burnham, 'From Avant-Garde to Specialism: Psycho-Analysis in America', *Journal of the History of the Behavioral Sciences*, XV (1979), p. 130.
p. 195 Quoted in S. Kraemer, '"The Dangers of This Atmosphere": A Quaker Connection in the Tavistock Clinic's Development', *History of the Human Sciences*, XXIV/2 (2011), p. 84.
p. 196 C. G. Jung, *Two Essays on Analytical Psychology*, in *The Collected Works of C. G. Jung* (Princeton, NJ, 1953–83), vol. VII, p. 117.
p. 197 S. Freud, letter to K. Abraham, 1908, in *A Psycho-Analytic Dialogue: The Letters of Sigmund Freud and Karl Abraham, 1907–1926*, ed. H. C. Abraham and E. L. Freud (London, 1965), p. 34.
p. 197 H. F. Ellenberger, *The Discovery of the Unconscious: The History and*

*Evolution of Dynamic Psychiatry* (London, 1970), pp. 695–6.

p. 198 C. G. Jung, 'Archetypes of the Collective Unconscious', in *The Archetypes and the Collective Unconscious*, *Collected Works*, vol. IX, part 1, p. 4.

p. 199 Jung, 'Archetypes of the Collective Unconscious', p. 40.

SEVEN: INDIVIDUALS AND SOCIETIES

p. 204 M. Bakhtin, 'Toward a Reworking of the Dostoevsky Book', in *Problems of Dostoevsky's Poetics*, trans. and ed. C. Emerson (Minneapolis, 1984), p. 287.

p. 204 Darwin, *Descent of Man*, vol. I, p. 84.

p. 204 K. Marx, *Economic and Philosophical Manuscripts*, trans. G. Benton, in *Early Writings*, ed. L. Colletti (London, 1995), p. 365.

p. 205 Allport, 'The Historical Background of Modern Social Psychology', p. 3.

p. 206 Broca, quoted in N. L. Stepan, 'Race and Gender: The Role of Analogy in Science', *Isis*, LXXVII (1986), p. 269.

p. 209 Bastian, quoted in K.-P. Koepping, *Adolf Bastian and the Psychic Unity of Mankind: The Foundations of Anthropology in Nineteenth Century Germany* (St Lucia, Queensland, 1983), p. 29.

p. 211 Boutmy, quoted in J. van Ginneken, *Crowds, Psychology, and Politics, 1871–1899* (Cambridge, 1992), p. 45.

p. 212 Le Bon, quoted in van Ginneken, *Crowds, Psychology, and Politics*, pp. 130–31.

p. 213 Tarde, quoted in van Ginneken, *Crowds, Psychology, and Politics*, p. 189.

p. 214 Baldwin, quoted in Allport, 'The Historical Background of Modern Social Psychology', p. 29.

p. 215 W. McDougall, *The Group Mind* (Cambridge, 1920), p. 7.

p. 217 F. H. Allport, *Social Psychology* (New York, 1975), p. 12.

p. 218 G. H. Mead, 'Social Psychology as Counterpart to Physiological Psychology', in *Selected Writings*, ed. A. J. Reck (Indianapolis, 1964), pp. 101–2.

p. 219 Mead, quoted in M. J. Deegan and J. S. Burger, 'George Herbert Mead and Social Reform', *Journal of the History of the Behavioral Sciences*, XIV (1978), p. 365.

p. 220 Barendregt, cited in Dehue, *Changing the Rules*, p. 111.

p. 226 Z. Barbu, *Problems of Historical Psychology* (London, 1960), p. 1.

p. 227 Quoted in Joravsky, *Russian Psychology*, p. 312.

p. 229 S. Toulmin, 'The Mozart of Psychology', *New York Review of Books*, XXV/14 (1978), pp. 51–7.

p. 229 D. Joravsky, 'L. S. Vygotskii: The Muffled Deity of Soviet Psychology', in *Psychology in Twentieth-Century Thought and Society*, ed. M. G. Ash and W. R. Woodward (Cambridge, 1987), pp. 189–211.

p. 230 Vygotsky, quoted in Joravsky, *Russian Psychology*, p. 260.

p. 230 A. R. Luria, *The Making of Mind: A Personal Account of Soviet Psychology*, ed. M. Cole and S. Cole (Cambridge, MA, 1979), p. 19.

p. 232 Rubinshtein, quoted in Joravsky, *Russian Psychology*, p. 372.

EIGHT: WHERE IS IT ALL GOING?

p. 238 J. Coulter, *Mind in Action* (Atlantic Highlands, NJ, 1989), p. 100.
p. 238 Cattell, quoted in Danziger, *Constructing the Subject*, p. 248.
p. 238 S. Koch, 'Introduction', in *A Century of Psychology as Science*, ed. Koch and Leary, p. 2.
p. 242 Boring, *History of Experimental Psychology*, p. 577.
p. 242 L. J. Cronbach, 'The Two Disciplines of Scientific Psychology', in *American Psychology in Historical Perspective: Addresses of the Presidents of the American Psychological Association*, ed. E. R. Hilgard (Washington, DC, 1978), pp. 435–58.
p. 250 K. Lorenz, *On Aggression*, trans. M. Latzke (London, 1966), p. x.
p. 251 Jensen, quoted in Fancher, *The Intelligence Men*, p. 194.
p. 252 E. O. Wilson, *On Human Nature* (Cambridge, MA, 1978), pp. x and 5.
p. 253 L. Cosmides, J. Tooby and J. H. Barkow, 'Introduction: Evolutionary Psychology and Conceptual Integration', in *The Adapted Mind: Evolutionary Psychology and the Generation of Culture*, ed. J. H. Barkow, L. Cosmides and J. Tooby (New York, 1992), p. 7.
p. 254 Yerkes, quoted in J. G. Morawski, 'Impossible Experiments and Practical Constructions: The Social Basis of Psychologists' Work', in *The Rise of Experimentation in American Psychology*, ed. J. G. Morawski (New Haven, CT, 1988), p. 81.
p. 256 C. Geertz, 'The Impact of the Concept of Culture on the Concept of Man', in *The Interpretation of Cultures: Selected Essays* (New York, 1973), pp. 49–50.
p. 256 J. D. George, 'The Decade of the Brain – A Decade of Scholarship: A Bibliography of the History of the Neurosciences, 1900–2000', *Journal of the History of the Neurosciences*, X (2001), pp. 113–21.
p. 257 S. Rose, *The 21st-Century Brain: Explaining, Mending and Manipulating the Mind* (London, 2006), p. 211.
p. 257 F. Crick, *The Astonishing Hypothesis: The Scientific Search for the Soul* (New York, 1994), p. 3.
p. 260 P. D. Kramer, *Listening to Prozac* (London, 1994), p. xiii.
p. 263 Rose, *The 21st-Century Brain*, p. 2.
p. 264 G. Lakoff and M. Johnson, *Philosophy in the Flesh: The Embodied Mind and Its Challenge to Western Thought* (New York, 1999), p. 5.
p. 266 C. A. van Peursen, *Body, Soul, Spirit: A Survey of the Body–Mind Problem*, trans. H. H. Hoskins (London, 1966), p. 166.
p. 266 A. J. Sutich, 'Introduction', *Journal of Humanistic Psychology*, I (1961), p. viii.
p. 269 C. Rogers, 'Toward a Science of the Person', in *Behaviorism and Phenomenology*, ed. Wann, p. 129.
p. 272 Nietzsche, *Beyond Good and Evil*, section 23.
p. 272 F. Nietzsche, *Ecce Homo*, trans. W. Kaufmann (New York, 1969), p. 266.
p. 272 F. Nietzsche, *Twilight of the Idols*, trans. R. J. Hollingdale (Harmondsworth, 1968), Foreword, p. 21.

p. 272 Nietzsche, *Beyond Good and Evil*, aphorism 68, p. 80; quoted by Freud, 'Notes upon a Case of Obsessional Neurosis', *Standard Edition*, vol. x, p. 184.

p. 275 R. Harré and P. F. Secord, *The Explanation of Social Behaviour* (Oxford, 1972), Preface.

p. 276 K. J. Gergen, 'Social Psychology as History', *Journal of Personality and Social Psychology*, XXVI (1973), p. 319.

p. 278 M. Weber, 'Science as a Vocation', trans. M. John, in *Max Weber's 'Science as a Vocation'*, ed. P. Lassman and I. Velody (London, 1989), p. 18.

# INDEX